THE BLUE ECONOMY

Concept, Constituents and Development

THE BLUE ECONOMY

Concept, Constituents and Development

Editors

Dr. Vijay Sakhuja

and

Cdr (Dr.) Kapil Narula

National Maritime Foundation

PENTAGON PRESS

The Blue Economy: Concept, Constituents and Development
Editors: Dr. Vijay Sakhuja and Cdr (Dr.) Kapil Narula

First Published in 2017

ISBN 978-93-86618-04-7

Published by
PENTAGON PRESS
206, Peacock Lane, Shahpur Jat
New Delhi-110049
Phones: 011-64706243, 26491568
Telefax: 011-26490600
email: rajan@pentagonpress.in
website: www.pentagonpress.in

Printed at Aegean Offset Printers, Greater Noida, U.P.

Contents

Foreword

The seas and the oceans around us are gaining new found importance as each day goes by because of their linkages with the 'Blue Economy' and there is no doubt that the current century is the "century of the seas". Our blue planet the Earth has a dominance of the maritime domain with over 70 percent of the Earth's surface covered by water, nearly 80 percent of the world population living within 200 nautical miles from the coast and about 90 percent of the world's trade transiting by sea. The Oceans are rich in oil and minerals resources, they are suppliers of oxygen, absorbers of carbon dioxide, a virtual heat sink, rich in bio-diversity, and have emerged as the global economic highways for transit of trade.

With depletion of resources on land, humankind has turned towards the oceans and there is a misperception that oceans have an unending resource base and are an infinite sink. Nothing could be further from reality. In the past few decades we have witnessed pollution of the oceans and contamination of the natural marine habitat with adverse impact of climate change on oceans. Studies have indicated that nearly 80 percent of the pollutants in the seas emanate from land and if the current rate of pollution continues, in a few decades we will have more plastic in the oceans than fish. Harnessing the ocean based Blue Economy therefore calls for efficient utilisation of marine resources without substantial environmental impact and ensuring sustained development of oceans.

India is essentially maritime nation with a natural out flow towards the seas, we have a long coastline of over 7500 km, an exclusive Economic Zone of over 2.0 million square kilometres, over 1300 islands and islets as a part of the Andaman and Nicobar group in the Bay of Bengal, Lakshadweep islands in the Arabian sea and along the West and East coast of India. We have 12 major ports and nearly 200 minor and intermediate ports, 90 percent of our trade by volume and 77 percent by value transits by sea. We have a vibrant shipbuilding industry and a thriving fishing industry. We have offshore oil and gas areas both off the West and East coast of India and deep sea bed mining interests in the Central Indian Ocean.

It is therefore evident that we have vast maritime interests which have a vital relationship with the nation's economic growth and are enablers of the Blue Economy. Honourable Prime Minister Shri Narendra Modi has outlined his maritime vision for the Indian Ocean through SAGAR which stands for 'Security and Growth for All in the Region'. The 'Sagarmala' project based on the four pillars of port modernisation, connectivity enhancement, port-led industrialisation and coastal community development has also been launched as a port-led development initiative. This will enable better connectivity of major and minor ports with the hinterland along with augmentation of coastal shipping and inland waterways. There is also a focus on maritime tourism and development of island infrastructure. There has therefore been a renewed thrust towards harnessing the Blue Economy in recent years and it appears as if India has once again turned towards the seas in her quest to emerge as a resurgent maritime nation.

In the last 10 years, the Indian Ocean Region has emerged as the world's centre of the gravity in the maritime domain. It is the third largest water body covering an area of 68.7 million sq.kms. and the countries on the ocean rim are home to one-third of world's humanity. The ocean is rich in oil and mineral resources and 66 percent of the world's oil, 50 percent of world's container traffic and 33 percent of world's cargo traffic transits over the waters of the ocean. Another distinguishing feature of the Indian Ocean is that 80 percent of the oil and trade which emanates in the region is extra regional in nature. This implies that if there is any impediment in the free flow of oil or trade, it would have a detrimental impact not just on economies of the region but global economies as well. It is therefore the collective responsibility of the navies and coast guards of maritime nations to ensure that we keep the global commons unpolluted, safe and secure. Networking among navies and coast guards and global maritime partnerships between maritime nations are therefore emerging as the new order of the 21st century.

This book discusses the concept of sustainable development, ocean resources, protection of maritime environment, and enablers of the Blue Economy. The chapters are academically rigorous and give insight into challenges and opportunities to harness the Blue Economy. They also provide recommendations for management of the multifarious activities in the maritime domain. I hope that this book adds to the discourse on the Blue Economy and helps to identify various avenues to partner together for sustainable development of the oceans.

Admiral RK Dhowan
PVSM, AVSM, YSM (Retd.)
Chairman
National Maritime Foundation

नितिन गडकरी
NITIN GADKARI

मंत्री
सड़क परिवहन राजमार्ग
एवं पोत परिवहन
भारत सरकार
MINISTER OF ROAD TRANSPORT
HIGHWAYS & SHIPPING
GOVERNMENT OF INDIA

सत्यमेव जयते

MESSAGE

India is a maritime nation and like other countries, India too is developing its maritime infrastructure to increase trade and reduce logistics costs. The National Maritime Foundation, through their Annual Maritime Power Conference on Blue Economy, has generated awareness among scholars and policy makers about the need for sustainable development of the seas and oceans.

I understand the importance of Blue Economy and we need to move ahead with transparent and positive attitude. Coasts are the growth engines of India's economy and so are the rivers. The maritime sector is moving fast, technologies are available, inland waterways and ports are developing, and financiers are ready to invest in tourism related infrastructure. It is therefore important that we implement a time-bound schedule for completion of various maritime projects, in a coordinated manner.

This book comprises many aspects of Blue Economy, which are essential for India's economic growth and is a valuable read for all stakeholders in the maritime industry. I appreciate the efforts of the National Maritime Foundations in highlighting various aspects of India's Blue Economy and the role which it can play in sustainable economic growth of the nation.

(Nitin Gadkari)

New Delhi
4th May, 2017

Transport Bhavan, 1 Sansad Marg, New Delhi-110 001. Tel.: 23711252, 23710121 (O), 23719023 (F)
Email : nitin.gadkari@nic.in

राजीव प्रताप रुडी
RAJIV PRATAP RUDY

सत्यमेव जयते

कौशल विकास और उद्यमशीलता
राज्य मंत्री (स्वतंत्र प्रभार)
भारत सरकार
Minister of State (Independent Charge)
Skill Development & Entrepreneurship
Government of India

June 19th, 2017

MESSAGE

Blue Economy of India, which has a great potential, still remains untapped. Keeping this in view, Government of India has been working forsustainable development of the oceans conforming to the principles of the Blue Economy. I sincerely consider that a partnership between skills and maritime domain is essential for promoting the Blue Economy. I also feel that there is a need to have an international conference on 'Skill Development in Maritime Economy', in collaboration with various stakeholders.

This book titled 'The Blue Economy: Concept, Constituents and Development'endeavors to cover various facets of the Blue Economy, its development and scope. It also throws light onhow India's potential for developing its Blue Economy can be nurtured. The compilation of papers in this book has been successfully authored by the participants of Annual Maritime Power Conference of the National Maritime Foundation held on 09 Feb 2017, which, I am sure, will go a long way in explaining the concept of Blue Economy and ultimately benefiting the common man.I am also confidentthat this bookwould contribute immensely to the development of Blue Economy in India and the region.

With my best wishes,

(RAJIV PRATAP RUDY

Skill India
कौशल भारत - कुशल भारत

Office: 2nd Floor, Shivaji Stadium Annexe, Shaheed Bhagat Singh Marg, New Delhi-110001
Tel. : +91-11-23450811, Fax. : +91-11-23450864
Residence: 22, Tughlak Crescent, New Delhi-110011, Ph. : +91-11-23070300, Fax : +91-11-23070999
@RajivPratapRudy, E-mail : rprudy.office@gmail.com, Website : www.skilldevelopment.gov.in

स्वच्छ भारत
एक कदम स्वच्छता की ओर

अमिताभ कांत
Amitabh Kant
ख्य कार्यकारी अधिकारी
Chief Executive Officer

सत्यमेव जयते

भारत सरकार
नीति आयोग, संसद मार्ग,
नई दिल्ली-110 001
Government of India
NATIONAL INSTITUTION FOR TRANSFORMING INDIA
NITI Aayog, Parliament Street,
New Delhi-110001
Tel. : 23096576, 23096574 Fax : 23096575
E-mail : ceo-niti@gov.in, amitabh.kant@nic.in

MESSAGE

I am delighted to note that the National Maritime Foundation which is a premier maritime think tank has now taken the lead role in promoting the Blue Economy. The era of Blue Economy is upon us, and India is trying to build on this concept. The NITI Aayog is preparing a 15-year vision document, where Blue Economy will form a major component, making it a part of India's policy framework.

The Indian Ocean region presents tremendous opportunities, both in terms of sustainable development and trade potential. The geostrategic position of India is vital and is aptly suited to develop a strong domestic and international trading network. The future of growth will come from oceans, and I believe that Blue Economy can serve as an indispensable platform for our country's development.

But the challenge is to learn from the mistakes of the West, and ensure that we do it in an integrated, holistic and sustainable manner. This would require education, mass communication and strong legislation to ensure that we do not do to the oceans what we have already done to the land.

I congratulate the National Maritime Foundation for the successful conduct of the Annual Maritime Power Conference and for the publication of this book which would richly contribute to the discourse on Blue Economy.

(AMITABH KANT)

Dated : 19.06.2017

Opening Address

Admiral Sunil Lanba, PVSM, AVSM, ADC, Indian Navy
Chief of the Naval Staff and Chairman Chiefs of Staff Committee

I am indeed honoured to be here at the inaugural session of this year's Annual Maritime Power Conference of the National Maritime Foundation and have the opportunity to address such a distinguished audience from India and overseas. The NMF has played a stellar role in enhancing maritime awareness within the country, and in moulding our strategic outlook on a wide range of maritime issues since its inception in 2005.[1] It has provided an excellent platform for interaction and helped in fostering a healthy exchange of ideas. This Annual Maritime Power Conference is an effort in that direction.

At the outset, my congratulations to the NMF for selecting an extremely relevant and forward-looking theme for this year's conference. The 'Blue Economy' is a concept that has far reaching implications, not only for India and the Indian Ocean Region, but the world at large. As resources over land get depleted, humankind will be forced to accord greater attention to the vast and largely untapped bounty of nature, in and below our oceans. However, as we move to explore the new frontiers, it is imperative that we remember our past mistakes, and do not end up leaving a trail of environmental destruction in pursuit of quick economic gains. This calls for a fresh approach – a new model of socio--economic development that integrates conservation and sustainability. And that is where the concept of 'Blue Economy' assumes great significance.

Blue Economy: The Concept

The 'Blue Economy' is a relatively new term. As we understand today, the

Blue Economy is a *"marine based economic development that leads to improved human well-being and social equity, while significantly reducing environmental risk and ecological scarcities".*[2]

It conceptualises oceans and their adjoining land areas as 'Development Spaces' where a sustainable use of resources is integrated into economic modelling and decision-making processes. However, the concept is still evolving and would continue to generate fresh ideas as it gets further debated.

The concept of the Blue Economy has attained a greater significance for a large number of countries in the Indian Ocean Region, since they have vast maritime zones under their jurisdiction. At the same time, there is a growing realisation that the business model for exploiting ocean-based resources must avoid a mindless pursuit of profit. Accordingly, a number of countries have included the Blue Economy in their national strategy.

India and the Concept of Blue Economy

Over the last few decades, India has embarked on an impressive growth trajectory. Concurrently, there is a growing realisation and acceptance within the country, of the strategic as well as economic value of the oceans.

Ladies and Gentlemen, with a coastline of more than 7,500 km and over 1,300 island territories, India is and has always been a maritime nation. Our renewed focus towards the oceans is evident in several initiatives of the Government of India. Its flag-ship initiative, *'Sagarmala',* which is aimed at harnessing the economic potential of the seas, has already identified more than 150 projects with an investment of approximately Rs 4 lakh crores.[3] When completed, these projects are expected to create more than 10 million jobs.

During our Hon'ble Prime Minister's maiden visit to the Island countries of the Indian Ocean in March 2015,[4] he underscored the need for harnessing the economic potential of the maritime commons. Our national vision is clearly articulated in the acronym SAGAR, which in the Hindi language means the 'ocean', and stands for 'Security and Growth for All in the Region'. In several ways, this approach signifies a secure, safe and sustainable model of development, with positive ramifications for the entire region. India has also been cooperating closely with a number of countries in the region such as Mauritius, Seychelles, Australia and Bangladesh to implement and harness the concepts of the Blue Economy.

The Role of Navy for a Conducive Maritime Environment

For any meaningful activity to be pursued, a secure and stable environment is a crucial prerequisite. I am glad to note that this conference will be discussing the subject of maritime security from the perspective of the Blue Economy. Maintaining a safe and conducive environment for an unhindered growth of ocean-based economic activities would require sustained efforts by the navies and maritime security agencies of the region. In a regional context, the sea lanes of the IOR are known to be the economic and energy highways that are critical for the global economy. As the strategic and economic value of these marine highways and regional marine resources increase, the multitude of challenges which confront us can also be expected to rise.To address these challenges, it is imperative that the navies and maritime agencies of the region join hands to anchor collective security, stability and sustained growth for the entire region.

The Indian Navy has always adopted a cooperative approach towards ensuring a secure maritime environment. At the request of our maritime neighbours, we are actively cooperating in the conduct of EEZ surveillance, hydrographic surveys, maintaining regional Maritime Domain Awareness as well as naval training and shipbuilding activities. The IONS initiative, which started in 2008, has also evolved into a key regional forum for maritime cooperation in the I0R. As we march forward, the Indian Navy will continue to accord a greater focus to enhance the collective ability of countries in the Indian Ocean to monitor and secure their maritime zones. I believe that it is our duty to apply maritime power so that we overcome our common security challenges and reinforce our rich economic and civilisational linkages.

Areas for Deliberations

As we discuss the constituents and avenues of Blue Economy during this conference, I would urge the distinguished gathering here to ponder over certain core issues which, in my opinion, merit greater attention. I would now briefly touch upon three such issues.

Firstly, there is a need for a suitable policy and legal framework at the national and international level to address different aspects of the Blue Economy. We also need to identify those existing national or international laws and policies which should be amended or modified to include tenets of the Blue Economy. For instance, we need suitable safeguards to be introduced for fisheries management so that the practice confirms to the tenets of the Blue Economy.

The second aspect pertains to the necessity of a collective approach. We need to explore ways and means of working closely and sharing expertise in areas such as fishing, shipping, offshore energy and coastal tourism, before gradually moving on to further technologically-intensive fields.

Thirdly, a science-based approach is essential for the development of the Blue Economy. It would also require considerable research in various ocean-based economic activities. A few countries, with their excellent scientific and technological base, are pioneers in this field. However, there are others which require hand-holding. This again goes to underpin my earlier point – the need for a very high degree of international cooperation if the concept of the Blue Economy has to take firm roots.

Conclusion

I am certain that it was with these thoughts that the subject for this Annual Maritime Conference was so aptly chosen by the NMF. I am sanguine that the participants will hold comprehensive deliberations on all issues related to the concept of the Blue Economy and explore avenues to develop a secure, sustainable model for economic development whilst ensuring ecological conservation. I am sure that the range of discussions over the next two days would be intellectually stimulating, thought-provoking and rewarding.

With these words, I wish the conference a resounding success. Thank you ladies and gentlemen, for your attention. Jai Hind.

REFERENCES AND NOTES

1. The NMF was launched on Feb 15, 2005 by Shri Pranab Mukherjee, the then Raksha Mantri.
2. Based on the 'approved' definition of Blue Economy at the "Rio+20" UN Conference on Sustainable Development (UNCSD), held in Rio de Janeiro in June 2012.
3. Pib.nic.in/newsite/printrelease.aspx?relid=137106.
4. PM Modi visited Mauritius, Seychelles and Sri Lanka in March 2015.

Introduction

Vijay Sakhuja and Kapil Narula

Oceans cover nearly 70 per cent of the earth's surface and provide us with a wide range of goods and services. They are an important source of food, energy, minerals and support livelihoods through shipping, shipbuilding and marine tourism industry. Critical ecosystem services such as being a supplier of oxygen, absorber of carbon dioxide, driver of global weather system, heat sink and biodiversity are provided by the oceans. They are therefore essential for life on earth and healthy oceans are vital for a thriving planet.

There is strong belief that the oceans are an inexhaustible resource and an infinite sink. This has resulted in a widespread degradation of ocean health and over the past few decades, there has been increasing evidence of overexploitation of fish, dumping of untreated pollutants from land into the sea, ship-borne marine pollution including oil spills, and contamination of natural marine habitat by alien species. The adverse impacts of climate change on the oceans are clearly visible and there has been an increase in acidification, warming and de-oxygenation affecting the aquatic life and marine biodiversity. Studies suggest that the ocean's ability to serve as a carbon and heat sink have thus reduced.

There are now calls for an efficient and sustainable use of sea-based resources and services without causing an environmental impact. This has led to the development of the discourse on the Blue Economy, which is now gaining momentum among the academia and policymakers. Accordingly, the theme of the Annual Maritime Power Conference - 2017 conducted by the National Maritime Foundation was chosen as ***"The Blue Economy: Concept, Constituents and Development"***. The conference offered a holistic view and identified several opportunities and challenges while proposing measures to implement the Blue Economy. This book emerges from the papers presented during the conference and supplements the understanding of the discourse on the Blue Economy.

The conference comprised five sessions. Session 1 deliberated on the concept of the Blue Economy and its contemporary relevance. It set the backdrop for the conference and discussed various interpretations such as the 'ocean economy' and the 'greening' of the ocean economy. The underlying principles, drivers and the means to implement the Blue Economy in the context of achieving sustainable development and balanced growth were deliberated. The key questions addressed during this session were:

- What is the difference between an 'ocean economy', 'coastal economy', and 'Blue Economy'?
- How does the Sustainable Development Goal (SDG) 14 for oceans fit into the Blue Economy?
- Why is international and regional cooperation essential for implementing the Blue Economy?

The second session on habitat protection and ocean ecology was devoted to the environmental aspects of the Blue Economy. Oceans are perhaps the least protected areas on the planet and habitat protection and marine ecology are critical but unstudied facets of the Blue Economy. The oceans provide animal and plant protein in the form of fisheries and seafood and are an important source of nutrition. However, indiscriminate fishing and destruction of marine biological resources has gradually lowered the capacity of the oceans to nurture marine life, leading to reduced ocean productivity. Marine Protected Areas (MPAs) and Integrated Coastal Zone Management (ICZM) are essential for enhancing the Blue Economy and Marine Spatial Planning is the key to ocean management. The questions that were deliberated during this session were:

- How can Illegal, Unreported and Unregulated (IUU) fishing be controlled and what measures can be implemented for the management of fish stocks?
- What are the challenges in implementing MPAs and ICZM and what steps can be taken for involving local communities?
- What is the role of 'ocean information systems' in protecting ocean ecology and how can they be leveraged for an effective monitoring and implementation of regulations?

The third session focused on ocean governance and marine resources. There are both legal as well as governance gaps in the Exclusive Economic Zone (EEZ) and in the Areas Beyond National Jurisdiction (ABNJ). The 'High Seas' are the 'common heritage of mankind' and even landlocked countries have equal rights and responsibilities in these areas. This session outlined the maritime regimes and the legal limitations in ocean governance including implementation gaps, which preclude the growth of the Blue Economy.

Offshore hydrocarbons are an important source of energy. Gas hydrates and natural gas in the seabed provide clean energy. On the other hand, marine renewable energy such as waves, tidal and offshore winds have large potential and can be harnessed to generate electricity as well as fresh water using appropriate technologies. Deep seabed mining holds immense promise for mineral resources such as polymetallic nodules and rare earths. Advances in marine bioprospecting and biotechnology are leading to the exploration of marine genetic resources and their derivative products such as RNAs, proteins, polysaccharides, lipids, peptides, etc., as potential ingredients for medicines and cosmetics. Although the extraction of these resources is essential for the benefit of humanity, there are associated environmental impacts. The questions that were addressed in this session were:

- What are the 'legal' and 'implementation' gaps for protecting marine habitat and ecology in the 'High Seas'? What are the current efforts towards strengthening ocean governance?
- What are the challenges in harnessing sea-based renewable and non-renewable energy and what policies can accelerate the sustainable development of ocean energy resources?
- How can marine genetic resources and minerals from deep seabed mining be shared for the common benefit of humanity?

Session 4 highlighted the aspects of maritime security, industry and maritime administration. Maritime security is vital for the development of the Blue Economy. The security of maritime infrastructure such as shipping, offshore oil platforms, and ports is essential given that these contribute to the goals of the Blue Economy. Likewise, prevention and control of maritime pollution is an important charter of the maritime security agencies.

Maritime transportation is a major contributor to the world GDP and is an essential component of the Blue Economy. The marine leisure industry provides alternate pathways to economic growth while providing employment;

however, there are many challenges. The issues addressed in the fourth session were:

- The role of maritime security and safety agencies and their ability to enforce laws and regulations to enhance the Blue Economy.
- What is the role of the shipping industry and how can it contribute to the Blue Economy?
- Can the marine leisure industry act as a catalyst to generate jobs and economic growth?

The last session discussed the enablers of the Blue Economy. Apart from various thematic issues and sectors of the Blue Economy discussed in the earlier sessions, there are common issues such as policies, funding, business models, innovation and technological developments which act as catalysts for enabling the Blue Economy. This session explored some important questions such as:

- How can the governments plan and finance programmes to enhance the Blue Economy? Can a public–private partnership model be developed?
- What is the role of technology as a facilitator and what are the necessary policies which would encourage the growth of the Blue Economy?
- What is the role of entrepreneurs and SMEs and how can innovations enhance the Blue Economy?

This book has attempted to capture the essence of the diverse components of the Blue Economy and provide an overview of these issues. The concept, thematic issues, regional as well as country perspectives and enablers of the Blue Economy are presented in thirteen separate chapters and the last chapter summarises the key issues and conference takeaways. We hope that this book adds to the body of existing literature on the Blue Economy and clarifies some of the concepts, discusses the constituents and outlines the developments in the Blue Economy in India and the world.

1

Blue Economy and Sustainable Development Goals: Aligned and Mutually Reinforcing Concepts

Kapil Narula

Oceans are a major source of economic activity and contribute to approximately five per cent of the world's GDP. The social and the environmental dimension are equally important and healthy oceans are essential for a healthy planet. Although the oceans provide humankind with its bounties, they have witnessed an extensive degradation over the last century. On the other hand, technological developments have now enabled undersea exploration and the growing demand of natural resources has put the fragile marine environment under tremendous pressure. In this scenario of increasing pursuit of oceanic resources while responding to the critical need for improving the health of oceans, the concept of the Blue Economy brings a ray of hope, as it intends to bring a paradigm shift in the way oceans and seas are utilised. The chapter traces the development of the discourse and highlights the concept of the Blue Economy. It distils the key characteristics of the Blue Economy and identifies the specific criterion which needs to be fulfilled so as to distinguish it from the ocean economy. The chapter thereafter focuses on the Sustainable Development Goals (SDGs) and draws the connections between the Blue Economy and the SDGs. Some challenges for the Blue Economy are discussed before concluding that the Blue Economy and SDGs are mutually aligned and reinforcing notions.

Introduction

Arthur C. Clarke, the famous English science and science fiction writer wrote, "How inappropriate to call this planet Earth when it is clearly Ocean?" With oceans covering 362 million square kilometres of the earth's surface (approximately 71 per cent of the total earth's surface), the above statement and the sobriquet for earth as the 'blue marble' are vindicated. Somewhat unknown is the fact that oceans provide 99 per cent of the earth's living space[1] and are the largest ecosystem on this planet.

Oceans are a major source of economic activity such as shipping, shipbuilding, ports, fishing, trade and offshore oil and gas production, etc., and contribute to approximately five per cent of the world's GDP. Eight of the top ten largest cities in the world are located along the coast and approximately 44 per cent of the world's population lives within 150 km of the coastline.[2] Over three billion people depend on marine and coastal resources for their livelihoods and more than 90 per cent of the world's trade is transported via the sea route. The World Wildlife Fund (WWF) in its report titled, "*Reviving the Ocean Economy: The case for action – 2015*",[3] reports that the value of the key ocean assets is conservatively estimated to be at least US$ 24 trillion and the annual value of goods and services contributed by the oceans is US$ 2.5 trillion.

The social dimension is equally important as the oceans contribute extensively to food security and fisheries are an important source of livelihood. Marine and coastal tourism is a large contributor to thriving cities and provides for recreational opportunities and cultural exchanges which are valuable for society. The environmental facet of the oceans is extremely relevant as they are the primary regulator of the global climate and act as a natural heat buffer. Further, the oceans are a large absorber of carbon dioxide and phytoplankton in the seas contribute to more than 50 per cent of the world's oxygen. Oceans are also critical for the protection and continuation of biodiversity and hence have a large intrinsic value. Therefore, development of the oceans in economic, social and environmental dimensions is essential for the well-being of the planet.

While oceans provide humankind with its bounties, they have witnessed extensive degradation over the last century. Although each threat differs in intensity and geographic distribution, together, they have lowered the ocean's overall health. The major threats include climate change leading to an increase in the heat content of the oceans, acidification and de-oxygenation.[4] These physical and chemical changes are leading to a loss of biodiversity and the destruction of corals. It is estimated that 20 per cent of the coral reefs have been completely lost globally and 75 per cent of the world's coral reefs are at risk from local and global stresses such as coral bleaching.[5] The destruction of marine habitat accompanied by a spread of invasive alien species and overfishing has also contributed to the lowering of ocean productivity. Approximately 53 per cent of the world's fisheries are fully exploited, and 32 per cent are overexploited, depleted, or recovering from depletion.[6] According to the FAO, global per capita fish consumption has increased from 22 pounds of fish per year in the 1960s to 44 pounds in 2015[7] leading to an overharvesting of fish at rates which are higher than the biologically sustainable levels. The danger of sea level rise has also resulted in increased levels of risks from flooding and inundation to coastal regions. This threat has increased considering that mangrove forests have been reduced to 30–50 per cent of their historical coverage and that 29 per cent of the sea grass habitat has disappeared.[8] Experts caution that a permanent damage might already have been done as many of these changes may be irreversible.[9]

On the other hand, technological developments have now enabled subsea exploration and a growing demand of natural resources have put oceans in the limelight. In this scenario of competing goals of increasing pursuit of oceanic resources, and the critical need for improving the health of oceans for the well-being of the planet, the concept of Blue Economy brings a ray of hope and intends to bring about a paradigm shift in the way the oceans and seas are utilised.

Development of Discourse on the Blue Economy

The term 'Blue Economy' was first used by author Gunter Pauli in 2004 and it was the philosophy of 'Zero Emissions Research and Initiatives' (ZERI) in action. This concept of the Blue Economy was inspired by nature's design principles where there is no 'waste product'. Based on this underlying principle of generating zero waste, the salient points of the Blue Economy were:[10]

(a) To respond to all basic needs of all living species with what is available
(b) Adopting a systemic approach and solving more than one problem at a time
(c) Converting the waste of one species into nutrients and energy for another species
(d) Based on a local system of production and consumption

(e) Adoption of innovative business model to bring products/services to the market and building social capital
(f) Embracing competitiveness and optimising values by connecting the untapped local potential

The first book titled *The Blue Economy* by Gunter Pauli was published in April 2010 and showcased 100 innovations which could create 100 million jobs. This was followed by its sequel titled *The Blue Economy-Version 2.0* which was released in May 2015 and highlighted 200 projects, investments of US$ 4 billion and 3 million jobs. The underlying principles of the Blue Economy continue to inspire entrepreneurs to create economic and social value without generating waste.

On a parallel front, the concept of Blue Economy developed in the context of sustainable development. The United Nations Conference on Environment and Development (UNCED), popularly known as the Earth Summit, was conducted in 1992 at Rio de Janeiro, Brazil. Apart from other successes such as the United Nations Framework Convention on Climate Change (UNFCCC) and the United Nations Convention on Biological Diversity (CBD), the outcome document of the summit, "Agenda 21", outlined the action plan of the UN and individual governments to attain sustainable development in the 21st Century and to minimise the human impact on the environment.[11] Twenty years later in 2012, the 'Rio+20' UN conference on sustainable development was held to reconcile the economic and environmental goals of development. Various practical measures for implementing sustainable development were outlined in its outcome document, "The Future We Want".[12] It was also decided to launch a process for developing a set of Sustainable Development Goals (SDGs), to replace the Millennium Development Goals (MDGs). The conference popularised the concept of the 'Green Economy' which had been developed by the United Nations Environment Programme (UNEP) and other organisations and proposed it as a tool to attain poverty eradication and sustainable development. The end goal of the 'Green Economy' was to "*...contribute to eradicating poverty as well as sustained economic growth, enhancing social inclusion, improving human welfare and creating opportunities for employment and decent work for all, while maintaining the healthy functioning of the earth's ecosystem*".

The relevance and applicability of the Green Economy was questioned by coastal countries and Small Island Developing States (SIDS) which had meagre land-based resources. In response to this need, the first Blue Economy

Summit was held in January 2014 at Abu Dhabi on the sidelines of the Abu Dhabi Sustainability Week under the aegis of the SIDS Action Platform. The summit succeeded in building an international momentum and in mainstreaming the concept of Blue Economy into the sustainable development agenda. The Blue Economy concept paper produced for the event and the 'Abu Dhabi Declaration' helped in reaching a common understanding of the term Blue Economy, which further developed over time.[13]

The Indian Ocean Rim Association (IORA) which is a regional organisation comprising of member states bordering the Indian Ocean also looked upon the Indian Ocean to promote sustainable growth and a balanced development of the region and outlined the principles for peaceful, productive and sustainable use of the Indian Ocean and its resources. This reaffirmation for the "protection, restoration, health, productivity and resilience of the Indian Ocean and its resources" was a key outcome of the Thirteenth Council of Ministers' Meeting (COMM) at Perth, Australia in November 2013.

The term 'Blue Economy' first appeared in the IORA Economic Declaration which was adopted by the Foreign Ministers of the member states of the IORA in October 2014 at Perth, Australia. The declaration stated that "*The Blue Economy – marine economic activity including fishing, renewable energy, mineral exploration and coastal tourism – is emerging as a common source of growth, innovation and job creation for the Indian Ocean region*", and these have become common priority areas for the countries of the IORA.[14]

Continuing this momentum, two events were conducted: the IORA Blue Economy Core Group Workshop on 'Promoting Fisheries & Aquaculture and Maritime Safety & Security Cooperation in Indian Ocean Region', Durban (South Africa), May 4–5, 2015 and The Indian Ocean Region Workshop on 'Exploration and Development of Seabed Minerals and Hydrocarbons: Current Capability and Emerging Science Needs', Bali (Indonesia), July 26–27, 2015. The first IORA Dialogue on the Blue Economy was held at Goa (India), August 17–18, 2015 where the key aspects of the Blue Economy were deliberated. These included an accounting framework; fisheries & aquaculture; renewable ocean energy; ports, shipping and manufacturing services and sea-bed explorations and minerals.

Following up on these meetings, the first IORA ministerial Blue Economy conference was held at Mauritius in September 2015 where the 'Mauritius Declaration on Blue Economy' was adopted.[15] Fisheries and aquaculture which contributed to food security, poverty alleviation and sustainable livelihoods;

renewable ocean energy to reduce the cost of energy and to mitigate and adapt to impact of climate change; seaport and shipping to promote trade, investment and maritime connectivity in the Indian Ocean rim region and, offshore hydrocarbons and seabed minerals to foster new business opportunities and attract investment in the Indian Ocean were adopted as the priority sectors of the Blue Economy. The declaration also recognised the importance of Sustainable Development Goals (SDGs) for the Blue Economy.

The SIDS initiative continued in 2016, the participating heads of state reconfirmed their commitment to the concept with the 2016 Abu Dhabi Declaration on the Blue Economy. It discussed the development and application of the Blue Economy approach to "*support the implementation of the Paris Agreement on Climate Change and the realisation of all relevant Sustainable Development Goals and particularly Goal 14 and its related targets*".[16]

Apart from the IORA, the Blue Economy and its variants have been adopted by various international organisations. The first Asia Pacific Economic Cooperation (APEC) Blue Economy Forum was held in 2011 at Xiamen, China, on the theme 'Promoting the Green Growth of Marine Economy'. This was debated and discussed over the next few years and in the fourth APEC Ocean-Related Ministerial Meeting in 2014, APEC acknowledged the importance of the Blue Economy and adopted the Xiamen declaration for the cooperation on the Blue Economy in the Asia-Pacific region. In the European Union, the 'Blue Growth strategy' which focused on jobs, value and sustainability was adopted and its role in the integrated maritime policy was acknowledged. The 'Limassol Declaration' 2012 also highlighted the marine and maritime agenda for growth and jobs and the importance of the Blue Economy for a smart, sustainable and inclusive growth. The African Union Foundation adopted the 2050 Africa's Integrated Maritime Strategy (AIM strategy) along with its theme 'Harnessing the Blue Economy in Achieving the African Union Agenda 2063' in 2015.[17] Partnerships in Environmental Management for the Seas of East Asia (PEMSEA), SAARC, Bangladesh, Seychelles, Maldives, Indonesia, Australia, Mauritius, India and other countries are also actively promoting the Blue Economy.

Characteristics of the Blue Economy

The term has caught the fancy of various countries and there are many interpretations of the concept of Blue Economy. The intelligence unit of *The Economist* adopts the following definition: "*The Blue Economy is one where economic activity is in balance with long term capacity of ocean ecosystems to support this activity and remain resilient and healthy*".[18] The IORA defines it as "*Integration of ocean economy development with the principles of social inclusion, environmental sustainability and innovative, dynamic business models*".[19] The APEC views the Blue Economy as "*an approach to advance sustainable management and conservation of ocean and coastal resources and ecosystems and sustainable development, in order to foster economic growth*".[20] As seen from the range of interpretations, there is no universally accepted consensus on the definition of the Blue Economy. However, there is a common understanding that it is marine-based economic development.

The concept of the Blue Economy emerged in response to the need of the SIDS but was anchored in the principles of the 'Green Economy'. The Green Economy is interpreted as "*one that results in improved human well-being and social equity, while significantly reducing environmental risks and ecological scarcities. It is low carbon, resource efficient and socially inclusive*".[21] Similarly, the Blue Economy is also a tool to address sustainable development challenges and poverty eradication but is better suited to the circumstances, constraints and challenges faced by Small Island Developing States (SIDS) which have smaller land-based resources.

The Blue Economy can be interpreted as a marine-based economy which has sustainable development as its end goal and where social-economic development is decoupled from environmental degradation. The desired outcome of a Blue Economy is social equity and improved human well-being. The Blue Economy is a source of innovation and job creation and it reduces ecological scarcity and environmental risks. The Blue Economy is based in a developing world context and focuses on the untapped resource potential of oceans. It considers the oceans as common heritage of humankind and focuses on the management of ocean resources. The Blue Economy views oceans as development spaces and attempts to balance conservation and resource extraction. The Blue Economy lays emphasis on resource efficiency where resource management and optimisation play a key role. The Blue Economy advocates that the monetary value of the oceanic ecosystem services should be included into economic modelling and decision-making processes. It also promotes steps to mainstream equity – in access to resources, development of resources and in the sharing of benefits. At the national level it envisages social equity, gender equity, inclusive growth and jobs for all; while at the international level, it promotes the interests and equal rights for all countries including the landlocked ones in the 'high seas'.[22]

The Blue Economy consists of the traditional sectors of the Ocean Economy such as the living and non-living resources consisting of fisheries, aquaculture, offshore oil and gas, ocean energy and seabed minerals; trade facilitated by shipping and its associated activities of shipbuilding and repair, port infrastructure and supporting services; marine and coastal tourism and a host of ancillary services. It also offers significant opportunities in the emerging sectors such as deep seabed mining, biotechnology and in harnessing marine generic resources. The WWF estimates that the Blue Economy is worth 24 trillion USD and while this economic potential is the focus of the erstwhile 'ocean economy' or 'marine economy', the Blue Economy advocates a balanced growth with equal importance on social, economic and environmental dimensions.

In order to distinguish between the Blue Economy and older variants of the economy, refer to Table 1 which differentiates between brown, green and blue economy.

Table 1. Different Colours of the Economy

	Brown	Green	Blue
End goal	Maximise economic growth	Balance economic, social and environmental growth	Decoupling of socio-economic development from environmental degradation
Scope/coverage	All	Land-based	Focus on SIDS/coastal states with oceanic resources
Resource use	Very high/Free	Less (3R's-Reduce, Recycle, Reuse)	Ocean-based resources/ Local
Waste generation	High pollution	Low	Zero
Environmental degradation	High (Oceans as waste dumps)	Waste discharge within carrying capacity of the ecosystem	Nil
Value of ecosystem services	Not included in national accounting system	Principle of "Polluter pays"	Incorporate into economic activities
Status	Failed??; Deeply ingrained	Limited success??; High cost/subsidy	???; Technological innovations

Further, an ocean-based economy can be classified as a blue economy if all the activities that are undertaken satisfy the following criterion:

(a) No overexploitation of resources

(b) Should not disturb the marine habitat or pose a risk to marine biodiversity
(c) Zero environmental impact
(d) Should aid restoration of the marine ecosystem

Sustainable Development Goals 2030

The UN Millennium Declaration 2000 adopted the Millennium Development Goals (MDGs) which were at the forefront of the world's development agenda from 2000 to 2015. These were intended to provide the basic standard of living to all human beings and were based on the values of human dignity, freedom, equality and respect for nature. The 8 goals and 18 targets were aggressively pursued by national governments and international agencies and were evaluated using 48 indicators. Towards the end of this period, it was noted that there was an uneven progress on the attainment of these goals and that inequalities still persisted.

In the furtherance of the original goals, the plan of action for Sustainable Development was adopted by the United Nations in September 2015 and was titled '*Transforming our World: The 2030 Agenda for Sustainable Development*'.[23] The end goal was human and societal well-being as well as a prosperous, sustainable and equitable world. It aimed at harmonising economic growth and was based on the principles of social inclusion and environmental protection. The focus was on the five Ps, viz. People, Planet, Prosperity, Peace and Partnership. 17 goals, 169 targets and 304 indicators were agreed to and adopted universally along with a shared responsibility by the international community to meet these goals. These goals were considered as integrated and indivisible and equal importance was laid to these goals. The list of the 17 agreed goals, known as SDGs, are placed in Appendix A.

Sustainable Development Goal 14 specifically deals with the oceans and states, "*Conserve and sustainably use the oceans, seas and marine resources for sustainable development*". It has 10 targets and specific indicators to evaluate its progress. The list of the targets and its accompanying indicators for SDG 14 are placed in Appendix B.

Blue Economy and SDGs: Aligned and Mutually Reinforcing Notions

The notion of the Blue Economy and the SDGs are aligned and mutually reinforcing. While on the one hand the SDG 14 mirrors the concept of the Blue Economy with sustainable development as the end goal, on the other hand, the Blue Economy supports the sustainable development agenda 2030 and contributes to various SDGs. The specific SDGs and the contribution of the Blue Economy to these goals are presented in Table 2.

Table 2. SDGs and Blue Economy

	SDGs	Blue Economy
1.	No poverty	Presents employment opportunities and promotes economic growth, thereby contributing to poverty alleviation.
2.	Zero hunger	Fisheries and aquaculture contribute to food security and are the primary source of protein for a large number of people.
3.	Good health and well-being	Healthy ecosystems promote good health and build resilience.
4.	Clean water and sanitation	Desalination of seawater can provide large quantities of freshwater to coastal areas and island states.
5.	Affordable and clean energy	Marine renewable energy has the second largest potential after solar energy, is clean and is abundantly available.
6.	Decent work and economic growth	New jobs related to the Blue Economy provide opportunities for economic growth. Sea-based resources also provide marginal and artisanal jobs for local communities.
7.	Industry, innovation and infrastructure	Ocean-led industry and new business models encourage the development of infrastructure.
8.	Reduced inequalities	Oceans are the common heritage of humankind and provide public goods, thereby contributing to lowering inequalities.
9.	Sustainable cities and communities	Leisure spaces and marinas can be developed around waterfronts in cities to promote sustainable cities.
10.	Responsible consumption and production	The Blue Economy with its principle of local consumption and production reduces the need for seaborne transportation of goods.
11.	Climate action	Blue Economy promotes the use of non-fossil fuels for shipping and advocates measures to lower the impact of climate change.
12.	Life on land	The Blue Economy advocates the protection of oceans from environmental degradation. Healthy oceans support life on land.
13.	Peace, justice and strong institutions	The Blue Economy provides opportunities for mutual cooperation and promotes international institutions.
14.	Partnerships for the Goals	The Blue Economy contributes to all SDGs and reinforces sustainable development.

This synergy between the Blue Economy and the SDGs was acknowledged and explicitly stated in the Abu Dhabi Blue Economy Declaration 2016, where the parties reiterated their commitment to a "*revitalized global partnership as a means to ensure the implementation of the 2030 Agenda and its SDGs, including Goal 14*".[24] The relationship between the SDGs and the Blue Economy can be visualised in Fig. 1.

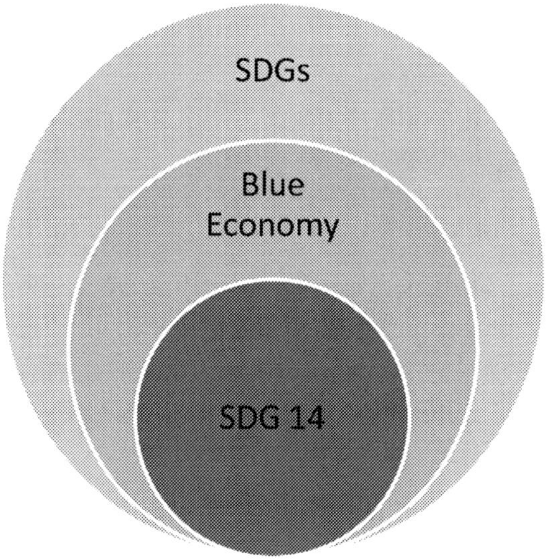

Figure 1. Relationship between SDG 14, Blue Economy and other SDGs

As is evident from Fig. 1, SDG 14 is *necessary* but not *sufficient* for the Blue Economy. Although SDG 14 is an essential part of the Blue Economy, the concept of Blue Economy goes much beyond SDG 14. Further, the Blue Economy aids the development of other SDGs and is essential to meet the remaining goals of the Sustainable Development Agenda 2030.

Challenges for the Blue Economy

Healthy oceans are essential for sustainable development and it is paramount that the focus shifts from purely economic growth to that of balanced growth where the environment is not seen as a hurdle but as a primary beneficiary of projects undertaken under the umbrella of the Blue Economy. Similarly, societal goals should be integrated during the design of the project or infrastructure development and should benefit the local communities directly or by incorporating measures which lead to inclusive growth. This is challenging as it involves a shift in the current way of thinking where economic growth is the primary objective of a majority of activities.

Some other challenges for the Blue Economy are as follows:

(a) Defining specific quantitative criteria for an activity so that it qualifies to be a part of the Blue Economy initiatives
(b) Coordinating the implementation of projects among different agencies
(c) Monitoring and tracking SDG 14 in the face of limited capabilities
(d) Financing activities under SDG 14 and providing trained human resources to support these activities
(e) Balancing the tradeoff between competing end-uses of oceanic spaces
(f) The risk of lower importance to SDG 14 vis-à-vis other SDGs

Some of these challenges can be overcome by adopting the following enablers of the Blue Economy:

(i) Integrated management: Efficient resource management is the key to Blue Economy. Marine Spatial Planning (MSP) and Integrated Coastal Zone Management (ICZM) and an efficient maritime administration are essential for harnessing the Blue Economy.
(ii) Technology: Continued research and development for ocean observation, innovative use of technology, collection of environmental and underwater data and integration of this data on accessible platforms is a key enabler of the Blue Economy.
(iii) Ocean governance: Well defined legal regimes for ocean exploitation, enforcement of regulations and maritime security are prerequisites for harnessing the Blue Economy.
(iv) Entrepreneurs: Financing the Blue Economy, supporting innovations during the implementation stage and flexible business models are also enablers of the Blue Economy.

Conclusion

The Blue Economy represents a paradigm shift in the way oceans and seas are utilised. It goes beyond the traditional thinking where oceans are looked upon only as resource providers and promotes oceans as life support systems which are central to the life on earth. Activities under the umbrella of the Blue Economy have to pass a stringent criteria which include zero environmental impact, leads to social inclusion and reduces environmental risks and resource shortages. The SDGs and the Blue Economy are fully aligned and the Blue Economy contributes to various other SDGs. While there are many challenges in the implementation of the Blue Economy, an integrated management of the oceans where the oceans are treated as development spaces, is the key enabler of the Blue Economy.

REFERENCES AND NOTES

1. Save the Sea, 'Interesting Ocean Facts'. http://savethesea.org/STS%20ocean_facts.htm (accessed Nov 10, 2016).
2. UN Atlas of the Oceans, 'Human Settlements on the Coast', http://www.oceansatlas.org/servlet/CDSServlet?status=ND0xODc3JjY9ZW4mMzM9KiYzNz1rb3M~ (accessed Nov 20, 2016).
3. Hoegh-Guldberg, O., The Boston Consulting Group: Douglas Beal, Taz Chaudhry, Hassan Elhaj, Amer Abdullat, Petra Etessy, Marty Smits, 'Reviving the Ocean Economy: The Case for action – 2015'. Gland: WWF International.
4. Narula, Kapil, 'Climate Change and Oceans: Growing Threat and Implications For India', *Maritime Affairs: Journal of the National Maritime Foundation of India*, 10 (2), 2014, pp. 96-119.
5. Burke, Lauretta Marie, *Reefs At Risk Revisited*, 1st edn, Washington, D.C.: World Resources Institute, 2011.
6. 'The State of World Fisheries and Aquaculture, 2010', 1st edn. Rome: Food and Agriculture Organization of the United Nations, 2010.
7. 'The State of World Fisheries and Aquaculture 2016: Contributing to Food Security and Nutrition for All', Rome: Food and Agriculture Organization of the United Nations, 2016.
8. Sustainable Development Knowledge Platform, 'Blue Economy Concept Paper'. https://sustainabledevelopment.un.org/. (accessed Nov 25, 2016).
9. *Science Daily*, 'Experts Call for Urgent Defence of Deep Ocean', May 15, 2014. http://www.sciencedaily.com/releases/2014/05/140515153947.htm (accessed Dec 03, 2016).
10. Zero Emissions Research and Initiatives,'The Blue Economy'. http://www.zeri.org/ (accessed Nov 25, 2016).
11. United Nations, 'Agenda 21', 1st edn. New York, 1992.
12. United Nations, 'The Future We Want?', New York, 2016.
13. Sustainable Development Knowledge Platform, 'Blue Economy, Abu Dhabi Declaration'. https://sustainabledevelopment.un.org/content/documents/2983BEdeclaration.pdf. (accessed Nov 25, 2016).
14. Indian Ocean Rim Association, 'IORA Economic Declaration'. http://www.iora.net/media/151280/iora_economic_declaration.pdf (accessed Dec 10, 2016).
15. Indian Ocean Rim Association, 'Mauritius Declaration on Blue Economy'. http://www.iora.net/media/158070/mauritius_blue_economy_declaration.pdf (accessed Dec 10, 2016).
16. Abu Dhabi Sustainability Week, 'Abu Dhabi 2016 Blue Economy Declaration'. http://abudhabisustainabilityweek.com/wp-content/uploads/2016/02/Abu-Dhabi-2016-Blue-Economy-Declaration-FINAL.pdf (accessed Dec 15, 2016).
17. African Union Foundation, 'Blue Economy'. http://www.africaunionfoundation.org/en/projects/blue-economy. (accessed Dec 15, 2016).
18. *The Economist*, Intelligence Unit, 'The Blue Economy: Growth, Opportunity and a Sustainable Ocean Economy'. https://www.eiuperspectives.economist.com/sites/default/files/images/Blue%20Economy_briefing%20paper_WOS2015.pdf (accessed Dec 15, 2016).
19. Indian Ocean Rim Association, 'Blue Economy'. http://www.iora.net/blue-economy/blue-economy.aspx (accessed Dec 15, 2016).

20. Ministry of Foreign Affairs of Japan, 'APEC Ocean Cooperation in the Asia Pacific Region', http://www.mofa.go.jp/mofaj/files/000059615.pdf (accessed Dec 15, 2016).
21. United Nations Environment, RIO+20 United Nations Conference on Sustainable Development, 'About Green Economy'. http://www.unep.org/rio20/About/GreenEconomy/tabid/101541/Default.aspx (accessed Dec 15, 2016).
22. 'High Seas' is a term used to refer to the area of the open ocean which is not a part of the exclusive economic zone, territorial sea or internal waters of any country.
23. UN General Assembly, *Transforming Our World: The 2030 Agenda for Sustainable Development*, 21 October 2015, A/RES/70/1, available at: http://www.refworld.org/docid/57b6e3e44.html (accessed 07 January 2017)
24. Abu Dhabi Sustainability Week, 'Abu Dhabi 2016 Blue Economy Declaration'. http://abudhabisustainabilityweek.com/wp-content/uploads/2016/02/Abu-Dhabi-2016-Blue-Economy-Declaration-FINAL.pdf (accessed Dec 15, 2016).

2

Role of Fisheries, Aquaculture and Management of Fish Stocks

Md. Khurshed Alam

Currently 3 billion people depend on fish for twenty per cent of their average per capita intake of animal protein. 97 per cent of the livelihoods that are directly dependent on fisheries and aquaculture occur in developing countries, mostly via small-scale operations in Asia. With 61.3 per cent of the world's fisheries now fully exploited, there is an urgent need to revise policies to ensure that the overexploitation and destruction of fisheries does not continue, and to deal with the problem of illegal fishing. Good governance is needed to ensure that fisheries and aquaculture adapt to the impacts of climate change and improve the resilience of food production systems. Therefore, a coherent and consistent regulatory framework and a strong national infrastructure will be required to facilitate the continuation of the existing sectors' contributions and the realisation of new opportunities. But the main challenge of the policy will be to improve the managerial and technical skill of the people, to develop community understanding and awareness of the general masses.

Introduction

The earth is distinguished from all other known planets by the presence of a warm and salty ocean that covers more than two-thirds of its surface. Eighty per cent of all life on the planet is found in the oceans. The ocean also provides a life support system that helps regulate climate and supplies half of the planet's oxygen. The fish contributes to nearly 20 per cent of the animal protein consumed by the world's population and thus is a critical source of food security, with the demand expected to double in the next twenty years.[1] Currently, 3 billion people depend on fish for 20 per cent of their average per capita intake of animal protein. 97 per cent of the livelihoods that are directly dependent on fisheries and aquaculture occur in developing countries, mostly via small-scale operations in Asia. The range of goods and services that flow from coastal and marine environments can be valued conservatively at US$ 2.5 trillion each year in addition to a growing range of products from the oceans, such as antibiotics, antifreeze, antifouling paints and a number of pharmaceutical products. The overall value of the oceans is approximately US$ 25 trillion.

The oceans also deliver essential public goods and services such as protection from natural hazards for the coastal population and carbon storage in the form of 'blue carbon' sinks such as mangrove forests, sea grass beds and other vegetated ocean habitats, which can sequester up to five times the amounts of carbon absorbed by tropical forests. Healthy oceans are also essential for global food security, livelihoods and are an engine for global economic growth.

The world faces one of the biggest challenges of the 21st Century: i.e. how to feed 9 billion people by 2050 in the face of climate change, economic and financial uncertainty, and the growing competition for natural resources. These multiple challenges require an integrated response and a transition of the world economy towards a sustainable, inclusive and resource efficient path. Ocean health and productivity must be at the heart of global efforts to eradicate poverty, strengthen food security and to build resistance to climate change.

Importance of the Oceans

The livelihoods and economic benefits derived from the oceans are valued by many ways especially in the form of the Blue Economy. The oceans are the primary means of trade and economic benefits from the oceans add greater value to natural endowments. A significant proportion of individual and collective wealth is derived from the oceans through shipping, seafood industry, tourism, biotechnology, petroleum and minerals, etc. Throughout the world, the movement of populations and economic activities are shifting to the coastal areas, and exploitation of oceanic resources and preservation of the marine environment are an increasing priority for states.

While the value of the oceans to our planet is incalculable, the fact of the matter is that the future of humanity is dependent on the health of the oceans and the goods and services they provide. Yet, this invaluable resource that feeds us, stabilizes the climate, and provides countless other benefits is showing serious signs of degradation due to local stresses such as overfishing, pollution, habitat destruction, as well as global phenomena of unprecedented changes in ocean temperature and acidity. With 61.3 per cent of the world's fisheries now fully exploited, and 28.8 per cent overexploited, depleted or recovering from depletion, there is a need to review the policies to ensure that overexploitation and destruction of fisheries do not continue. Further, habitat protection and fisheries management must be promoted with the aim of becoming ecologically sustainable.

Role of Fisheries

Fisheries and aquaculture is an important source of food, nutrition, income and livelihoods for millions of people around the world. The world per capita fish supply reached a new high of 20 kg in 2014, due to a vigorous growth in aquaculture, which now provides half of all fish for human consumption. There has also been slight improvement in the state of certain fish stocks due to improved fisheries management. Moreover, fish continue to be one of the most-traded food commodities worldwide with more than half of fish exports, by value, originating in developing countries. The recent reports by experts, international organisations, industry and civil society representatives all highlight the tremendous potential of the oceans and inland waters to contribute significantly to food security and adequate nutrition for a global population.

In 2013, the fish accounted for about 17 per cent of the global population's intake of animal protein and 6.7 per cent of all protein consumed. Moreover, the fish provided more than 3.1 billion people with almost 20 per cent of their average per capita intake of animal protein. In addition to being a rich source of easily digested, high-quality proteins containing all essential amino acids, fish provides essential fats (e.g. long-chain omega-3 fatty acids), vitamins (D, A and B) and minerals (including calcium, iodine, zinc, iron and selenium), particularly if eaten whole. Even small quantities of fish can have a significant positive nutritional impact on plant-based diets, and this is the case in many Low Income Food Deficit Countries (LIFDCs) and least-developed countries. Fish is usually high in unsaturated fat and provides health benefits in protection against cardiovascular diseases. It also aids foetal and infant development of the brain and nervous system. With its valuable nutritional properties, it can also play a major role in correcting unbalanced diets and, through substitution, in countering obesity.

The share of world fish production utilised for direct human consumption has increased significantly in recent decades, up from 67 per cent in the 1960s to 87 per cent, or more than 146 million tonnes, in 2014. The remaining 21 million tonnes were destined for non-food products, of which 76 per cent was reduced to fishmeal and fish oil in 2014, the rest being largely utilised for a variety of purposes including as raw material for direct feeding in aquaculture. Increasingly, the utilisation of by-products is becoming an important industry, with a growing focus on their handling in a controlled, safe and hygienic way, thereby also reducing waste.

In 2014, 46 per cent (67 million tonnes) of the fish for direct human consumption was in the form of live, fresh or chilled fish, which in some markets are the most preferred and highly priced forms. The rest of the production for edible purposes was in different processed forms, with about 12 per cent (17 million tonnes) in dried, salted, smoked or other cured forms, 13 per cent (19 million tonnes) in prepared and preserved forms, and 30 per cent (about 44 million tonnes) in frozen form. Freezing is the main method of processing fish for human consumption, and it accounted for 55 per cent of the total processed fish for human consumption and 26 per cent of total fish production in 2014.

Fishmeal and fish oil are still considered as the most nutritious and digestible ingredients for farmed-fish feeds. To offset their high prices, as the feed demand increases, the amount of fishmeal and fish oil used in compound feeds for aquaculture has shown a clear downward trend, with their being more selectively used as strategic ingredients at lower concentrations and for specific stages of production, particularly hatchery, bloodstocks and finishing diets.

International trade plays a major role in the fisheries and aquaculture sector as an employment creator, food supplier, income generator, and contributor to economic growth and development, as well as to food and nutrition security. Fish and fishery products represent one of the most-traded segments of the world food sector, with about 78 per cent of seafood products estimated to be exposed to international trade competition. For many countries and for numerous coastal and riverine regions, exports of fish and fishery products are essential to their economies, accounting for more than 40 per cent of the total value of traded commodities in some island countries, and globally representing more than 9 per cent of the total agricultural exports and 1 per cent of the world merchandise trade in value terms. Trade in fish and fishery products has expanded considerably in recent decades, fuelled by growing fishery production and driven by high demand, with the fisheries sector operating in an increasingly globalised environment. In addition, there is an important trade in fisheries services.

Capture Fisheries and Aquaculture

Global total captured fishery production in 2014 was 93.4 million tonnes, of which 81.5 million tonnes were from marine waters and 11.9 million tonnes were from inland waters. For marine fisheries production, China remained the major producer followed by Indonesia, the United States of America and the Russian Federation. The world catches in inland waters were about 11.9 million tonnes in 2014, continuing a positive trend that has resulted in a 37 per cent increase in the last decade. Sixteen countries have annual inland water catches exceeding 200,000 tonnes, and together they represent 80 per cent of the world total. The production of aquatic animals from aquaculture in 2014 amounted to 73.8 million tonnes, with an estimated first-sale value of US$ 160.2 billion. This total comprised 49.8 million tonnes of finfish (US$ 99.2 billion), 16.1 million tonnes of molluscs (US$ 19 billion), 6.9 million tonnes of crustaceans (US$ 36.2 billion) and 7.3 million tonnes of other aquatic animals including amphibians (US$ 3.7 billion). China accounted for 45.5 million tonnes in 2014, or more than 60 per cent of the global fish production from aquaculture. Other major producers were India, Vietnam, Bangladesh and Egypt. In addition, 27.3 million tonnes of aquatic plants (US$ 5.6 billion) were cultured. Aquatic plant farming, overwhelmingly of seaweeds, has been growing rapidly and is now practised in about 50 countries. Importantly in terms of food security and the environment, about half of the world's aquaculture production of animals and plants comes from non-fed species. These species include silver and bighead carps, filter-feeding animal species (e.g. bivalve molluscs) and seaweeds. However, growth in production has been faster for fed species than for non-fed species.

Fishers and Fishing Vessels

An estimated 56.6 million people were engaged in the primary sector of capture fisheries and aquaculture in 2014, of whom 36 per cent were engaged full time, 23 percent part time, and the remainder were either occasional fishers or of unspecified status. Following a long upward trend, numbers have remained relatively stable since 2010, while the proportion of these workers engaged in aquaculture increased from 17 per cent in 1990 to 33 per cent in 2014. In 2014, 84 per cent of the global population engaged in the fisheries and aquaculture sector was in Asia, followed by Africa (10 per cent) and Latin America and the Caribbean (4 per cent). Of the 18 million people engaged in fish farming, 94 per cent were in Asia. Women accounted for 19 per cent of all people directly engaged in the primary sector in 2014, but when the secondary sector (e.g. processing, trading) is included, women make up about half of the workforce.

The total number of fishing vessels in the world in 2014 is estimated at about 4.6 million, very close to the figure for 2012. The fleet in Asia was the largest, consisting of 3.5 million vessels and accounting for 75 per cent of the global fleet, followed by Africa (15 per cent), Latin America and the Caribbean (6 per cent), North America (2 per cent) and Europe (2 per cent). Globally, 64 per cent of reported fishing vessels were engine-powered in 2014, of which 80 percent were in Asia, with the remaining regions, under 10 per cent each. In 2014, about 85 percent of the world's motorised fishing vessels were less than 12 m in length overall, and these small vessels dominated in all regions. The estimated number of fishing vessels of 24 m and longer operating in marine waters in 2014 was about 64,000, the same as in 2012.The state of the world's marine fish stocks has not improved overall, despite a notable progress in some areas.

The State of Fishing

Based on FAO's analysis of assessed commercial fish stocks, the share of fish stocks within biologically sustainable levels decreased from 90 per cent in 1974 to 68.6 per cent in 2013. Thus, 31.4 per cent of fish stocks were estimated as fished at a biologically unsustainable level and therefore overfished. Of the total number of stocks assessed in 2013, fully fished stocks accounted for 58.1 per cent and under fished stocks 10.5 per cent. The under-fished stocks decreased almost continuously from 1974 to 2013, but the fully fished stocks decreased from 1974 to 1989, and then increased to 58.1 per cent in 2013. Correspondingly, the percentage of stocks fished at biologically unsustainable levels increased, especially in the late 1970s and 1980s, from 10 per cent in 1974 to 26 per cent in 1989. After 1990, the number of stocks fished at unsustainable levels continued to increase, albeit more slowly. The ten most-productive species accounted for about 27 per cent of the world's marine capture fisheries production in 2013. However, most of their stocks are fully fished with no potential for increases in production; the remainder is overfished with increases in their production only possible after successful stock restoration.

Overfishing, formally defined as "situations where one or more fish stocks are reduced below predefined levels of acceptance by fishing activities", means that fish stocks are depleted to the point where they may not be able to recover. In some cases, depleted fish stocks have been restored; however, this is only possible when the species' ecosystem remains intact. If the species depletion causes an imbalance in the ecosystem, not only is it difficult for the depleted stocks to return to sustainable levels, other species dependent on the depleted stocks may become unbalanced, causing further problems towards marine ecology. Overfishing, i.e. catching more fish than the ocean can produce, has been an ongoing challenge for decades for fisheries managers.

Related to overfishing is the question of how we catch fish. Certain types of fishing methods destroy or damage the seafloor habitats where fish and many other animals reside. Certain fishing methods are notorious for catching large amounts of by-catch, sea turtles, seabirds and marine mammals that are unintentionally caught and often intentionally killed in fishing operations. Among all fishing methods, bottom trawling, a fishing method that drags a large net across the seafloor is the most destructive fishing method which is common in bays and seas. Bottom trawling is an industrial fishing method where a large net with heavy weights is dragged across the seafloor, scooping up everything in its path from the targeted fish to the incidentally caught centuries-old corals. Bottom trawls are used in catching marine life that lived on the seafloor, like shrimp, cod, sole and flounder. Bottom trawls are also commonly used by other fishing nations and on the bays/high seas. Bottom trawling is unselective and severely damaging to the benthic ecosystems. The net indiscriminately catches every life and object it encounters. Thus many varieties end up being mistakenly caught and thrown overboard, dead or dying, including endangered fish. This collateral damage, called as by-catch, can amount to 90 per cent of a trawler's total catch. In addition, the weight and width of a bottom trawl can destroy large areas of seafloor habitats that give marine species food and shelter. Such habitats destructions can leave the marine ecosystem permanently damaged.

Fish Trading

Fishery trade represents a significant source of foreign exchange earnings for many developing countries, in addition to its important role in income generation, employment, food security and nutrition. In 2014, fishery exports from the developing countries were valued at US$ 80 billion, the net-export revenues (exports minus imports) reached US$ 42 billion; higher than other major agricultural commodities (such as meat, tobacco, rice and sugar) combined. China is the primary fish producer and largest exporter of fish and fishery products. It is also a major importer due to outsourcing of processing from other countries as well as a growing domestic consumption of species not produced locally. However, in 2015, after years of sustained increases, its fishery trade experienced a slowdown with a reduction in the fish-processing sector. Norway, the second major exporter, posted record export values in 2015. In 2014, Vietnam became the third major exporter, overtaking Thailand, which has experienced a substantial decline in exports since 2013, mainly linked to reduced shrimp production due to outbreak of disease. In 2014 and 2015, the European Union (member countries) was by far the largest single market for fish imports, followed by the United States of America and Japan. Developing economies, whose exports represented just 37 per cent of the world trade in 1976, saw their share rise to 54 per cent of the total fishery export value and 60 per cent of the quantity (weight) by 2014.

Management of Fisheries Stocks

Governance of fisheries and aquaculture should be greatly influenced by the 2030 Agenda for Sustainable Development, the Sustainable Development Goals (SDGs), and the Paris Agreement of the Conference of the Parties (COP21) of the United Nations Framework Convention on Climate Change (UNFCCC). The 17 SDGs and their 169 targets provide a framework to guide development actions of governments, international agencies, civil society and other institutions over the next 15 years with the ambitious aim of eradicating extreme poverty and hunger. Food security and nutrition, and sustainable management and use of natural resources, feature prominently in the SDGs and targets, applying to all countries, and integrating the three dimensions of sustainable development – economic, social and environmental.

Moreover, the Paris Agreement recognises that climate change is a fundamental threat to global food security, sustainable development and poverty eradication.

Thus, governments need to ensure that fisheries and aquaculture adapt to the impacts of climate change and improve the resilience of food production systems. FAO's 'Blue Growth Initiative' assists countries in developing and implementing the new global agenda in relation to sustainable capture of fisheries and aquaculture, livelihoods and food systems, and economic growth from aquatic ecosystem services. It promotes the implementation of the Code of Conduct for Responsible Fisheries (the Code) and the ecosystem approach to fisheries and aquaculture.

Reflecting the objectives of several SDGs, it especially targets the many vulnerable coastal and fisheries-dependent communities where ecosystems are already under stress from pollution, habitat degradation, overfishing and harmful practices. There is a need to strengthen the aquatic ecosystem governance to deal with the increasing use of water space and resources. It is necessary to coordinate the various activities taking place in a given region, recognise their cumulative impacts, and harmonise the sustainability goals and legal frameworks. This requires adding a layer of governance to deal with coordination across sectors and to ensure that common goals of environmental protection, ecosystem and biodiversity conservation are met while addressing social and economic development goals.

There is an urgent need to ban bottom trawling and adopt safe and sustainable methods of fishing to save the marine ecosystem. To protect the ocean ecosystems from the impacts of bottom trawling, the fishing operation should be switched over to other sustainable methods. These should be institutionalised from an in-depth research on fishing gear and only those methods which will be profitable in terms of fish ecology, commercial viability, and environment-friendly practices should be adopted. Partnerships can be very effective in improving the sustainability of fisheries and aquaculture.

Focusing on tuna and deep-sea fisheries, with an emphasis on enhancing global and regional coordination on the ABNJ (Areas Beyond National Jurisdiction) issues is also required. The Common Oceans ABNJ Program aims to promote an efficient and sustainable management of fisheries resources and biodiversity conservation in the ABNJ to achieve the internationally agreed global targets. The innovative five-year ABNJ Program, which started in 2014, is funded by the Global Environment Facility (GEF) and coordinated by FAO in close collaboration with three other GEF implementing agencies and a variety of partners. Another partnership initiative is the Global Aquaculture Advancement Partnership (GAAP) programme established by FAO. Its aim is to bring partners together to channel their technical, institutional and financial resources effectively and efficiently in support of global, regional and national aquaculture initiatives. Specifically, the GAAP seeks to promote and enhance strategic partnerships, and to use them to gather resources to develop and implement projects at various levels.

Implementing the ecosystem-based management into fisheries management also requires a careful account of the ways the fishing activities may affect marine ecosystems. In order to maintain a healthy ecosystem, and hence healthy fisheries, in many instances, the fish catches may need to be reduced to allow the target species to continue to play a natural role in functioning ecosystems. In addition, the fishing gear that eliminates or minimises the incidental offtake of non-targeted species and damage to marine habitats may need to be implemented. Marine ecosystems are very complex, its understanding is limited, and the ways in which the human activities affect them is poorly addressed. The ecosystem-based management approach to managing fisheries recognises that decisions will often be made in a climate of uncertainty. However, uncertainty should never be an excuse for inaction. The management decisions are best made using multiple lines of evidence, including convincing unreliable evidence and a precautionary approach. One key aspect of ecosystem-based management is that it recognises the economic, social and cultural interest of all stakeholders in fisheries and how these interests affect resource management. Therefore, attempts for adopting ecosystem-based management tools including a precautionary approach may be considered in the near future.

By managing human issues and impacts, the ecosystem-based management approach to fisheries is more likely to succeed where many other initiatives have failed – leading to healthy, sustainable fisheries and restored marine ecosystems. Marine Protected Areas (MPAs) is one of the simplest of marine management tools that can offer some of the highest gains in the terms of marine conservation. Given the logistical and economical challenges of monitoring a large marine ecosystem, huge expenses for monitoring and surveillance is one of the major challenges of the ocean, which threatens the achievement of public compliance to regulations on effective protection. New surveillance technologies can increase the monitoring capability and guide the law enforcing agencies towards suspicious activities. The surveillance and enforcement of regulations on the high seas is even more challenging.

Conclusion

Future development prospects of aquaculture appear promising. Well-managed coastal aquaculture and mariculture offer significant scope for green growth and employment opportunities for coastal communities at low levels of CO_2 emissions when compared to other protein production systems. Extracting wild natural resources (essentially fish) for animal consumption could be done and used by agriculture and aquaculture. The momentum of fast increasing aquaculture (by 160 per cent over the last decade) and with its huge success inland, it is important to replicate such expertise in sea aquaculture for fish, seaweed, pearl and oyster farming. Strengthening regional fisheries bodies, national fisheries management agencies, the fishing community and fish workers' organisations and private sector associations is critical to the sustainable and equitable use of marine resources through aquaculture. A comprehensive policy will reflect and be responsive to the inter-connections between the air, sea and land and to the physical and biological dynamics of the ocean and along the coast.

Ocean living resources, the ecosystems they form, and the ecological processes they mediate, provide immense benefits to human society. Increasingly, these resources and their benefits are threatened by human activities. Fishing, waste disposal and runoff, coastal development, and invasions of exotic species have led to worldwide declines in marine species and the integrity of marine ecosystems, particularly coastal ecosystems. The issues raised are at the heart of the shared concern for our marine environment. The development of the policy will enable utilisation, conservation and protection of the oceans. It can draw on the marine industry development strategy and can support marine science and technology plans.

A coherent and consistent regulatory framework and a strong national infrastructure are required to facilitate the continuation of the existing sectors' contributions and the realisation of new opportunities. These measures need to be pursued by all maritime agencies and stakeholders across the various sectors consistent with planning and management principles. Hence, all maritime sectors and industries like fisheries, aquaculture, offshore hydrocarbons, environment, shipping and shipbuilding, port development and tourism would have to provide positive responses against the challenges.

But the main challenge of the policy will be to improve the managerial and technical skill of the people, to develop community understanding and awareness of the general masses. Since the policy will have strong international connotations, the guidelines should also be established by identifying the common areas for regional cooperation with neighbours. The governments should formulate an Ocean Act before the problems of the ocean become irrecoverable and before missing any other opportunities that might put a heavy toll on the national economies.

REFERENCES AND NOTES

1. The State of World Fishes and Aquaculture, FAO, Rome 2016

3

Role of Marine Energy in Tropical Waters to Support Regional Growth and Energy Security

Narasimalu Srikanth

Ocean energy systems form a viable clean energy source from waves, tides, ocean current and ocean thermal gradient, etc,. and are well suited to support the energy needs of countries in the tropical region. However, barriers towards their adoption as an alternative energy source includes their presumed high cost of energy, misfit towards regional needs, lack of regional supply chain, environmental concerns due to perceived interference with the ecosystem and other marine industries. However, governments can identify remote coastal regions as potential adopters of support basic energy needs including desalination, apart from a generation of electricity. This chapter discusses the ocean energy choices, the associated resource mapping and harvesting technologies and the environmental impact assessment of ecological and aquaculture receptors, viz., water quality, seabed geology, underwater acoustics and hydrodynamics to study the impact on the marine industries. Further it highlights the technology hybridisation to tune ocean renewables to meet regional energy needs, to co-evolve the necessary supply chain and to prepare the domestic market for accelerating the adoption of ocean energy for supporting the local economy and job creation.

Introduction

Ocean energy is an emerging sector and is new to the maritime space. It is possible to extract energy from waves, tides, ocean current and ocean thermal gradient, etc., and these have contributed towards meeting Europe and North America's renewable energy goals. The World Energy Council considers that 0.1 percent of the ocean energy can suffice to meet all human energy needs.

Regional growth depends on various marine industries that simultaneously use the ocean space viz., shipping, offshore oil and gas, defence, tourism, fishing and aquaculture, etc., and they have a common concern that the deployment of the ocean energy system may have an impact on the ecology of the ocean. Survey shows that decision-makers lack awareness about energy resource potential in the oceans, the perceived high cost of energy and the fear of environmental threat towards damaging the marine ecology and other marine industries which pose as barriers towards ocean energy adoption.

This chapter suggests that the adoption of Ocean Energy Systems (OES) towards remote coastal and islandic regions will be a good start to provide the necessary confidence towards a mainstream adoption of the OES. Attempt is also made to introduce the essential tasks in a systematic Environmental Impact Assessment (EIA) study that could provide the necessary baseline data to monitor and take the further necessary actions during an OES deployment and operation. To minimise uncertainties, the chapter suggests promoting inter-industry learning from other similar technological industries to follow the best practices for minimising the environmental impact. The environmental impacts of the OES in tropical waters is discussed in terms of necessary key receptors that are to be identified for specific sites, to define the individual components of the environment such as flora, fauna, soil, water, air, climatic factors, sound, seabed landscape and their interactions.

Resource mapping for a better system deployment and optimum OES sizing, in tune with the energy resource availability, is the key that can help minimise the cost of electricity generation and can enhance the energy conversion potential. The resource mapping practices specifically towards wave, tidal and ocean thermal energy are described in a separate section. Based on the resource understanding, device-site matching methods are to be adopted to choose the appropriate OES to maximise the capacity factor of the device and to achieve the optimum techno-economics in a project.

Further, to overcome the barriers towards OES adoption and to promote regional growth, indigenisation towards modifying the OES for meeting regional needs such as in remote coastal zones is to be focused and further, the system is to be tuned towards specific applications such as desalination. Hence the co-creation of technology indigenisation and supply chain creation helps to accelerate the mainstream adoption of the OES as an alternative energy source.

Ocean Energy Harvesting

Ocean renewable energy is a promising energy source, and has attracted considerable amount of interest in recent times. The oceans have tremendous amounts of energy such as wave energy, tidal energy, thermal gradient along the ocean depth, salinity gradient, etc., and these if harvested, can benefit the remote coastal locations and the island communities where the population density is high and purely supported through by fossil fuels. Various ocean energy forms listed in Table 1, mainly originate from the thermal energy provided by the sun and from mechanical energy from the lunar gravity.

Table 1. Ocean Renewable Energy Resources: Types and Global Potential

Energy Resource	Energy Potential (Twh/Yr)
Thermal gradient (OTEC)	10,000
Waves	80,000
Tidal Range (Barrage)	300+
Tidal/Marine Current	800+
Salinity Gradient	2000

Tidal barrages are technologically mature projects. In 2012, there was more than a 500 MW installed capacity worldwide, mainly in two projects in South Korea and France. Although mature, it is controversial because of the effect on estuary landscape and ecology as well as the effects on animals. More objections are surfacing from governments and environmentalists due to this and the forecast of world tidal barrage capacity appear to be dampening. The tidal barrage in La Rance Barrage (France) and Shinwa Barrage (Korea) have a combined capability of 500 MW.

The tidal stream energy offers enough confidence as an energy source due to its high predictability. There are a variety of novel device concepts being developed globally but these needs to be assessed in the field after lab evaluation. These designs should converge to one or few dominant designs ready to be adopted, as observed in the onshore wind industry where a three-bladed horizontal axis wind turbine is the design of choice within the industry. Hence, test bedding under a well observed field condition is needed to characterise the device to understand the capacity factor, asset cost, operating cost and cost of electricity generation. However, designs are being evolved for new generations and new markets such as in tidal turbines. The first-generation turbines were being designed as seabed-mounted, while in the second generation, semi-buoyant and totally floating designs are being pursued with a high degree of product modularity and visual impact as illustrated in Fig. 1 to counter the environmental impact challenges and minimal operating costs.

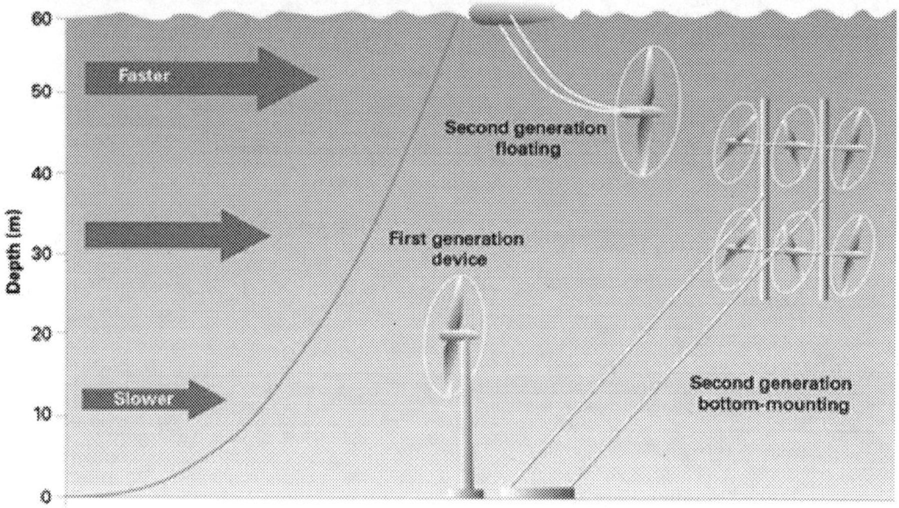

Figure 1. Different types of tidal energy turbines

Similarly, wave energy is a promising source as it is predominantly caused by flowing winds on the sea surface that can be converted into electricity by a wave energy converter (WEC). Many onshore projects, such as the Pico Island plant and the Islay plant are operating in the Azores and Scotland, respectively. In the same way as tidal technologies, various wave power projects have been implemented in the past two years. In the United Kingdom, 14 wave devices totalling 3.6 MW were deployed in 2012 for testing by the EMEC. In South Korea, the Yongsoo 500 kW oscillating water column (OWC) pilot plant at the Jeju Island is expected to start operation. In Norway, the Lifesaver, a 16-meter-diameter point absorber concept developed by Fred Olsen, has been undergoing sea trials off the coast of Cornwall in the UK, with a total installed capacity of 400 kW. In Denmark, Wavestar is currently testing its prototype with two floaters and the maximum production measured to date is 39 kW. In Portugal, the WaveRoller prototype (300 kW) was installed in the summer of 2012 and is connected to the grid. In Spain, two wave energy demonstration projects are progressing: the Mutriku OWC plant (200 MWh of energy production after one year of operation) and the WELCOME—Wave Energy Lift Converter Multiple España (150 kW prototype at 1: 5 scale) developed by PIPO Systems, installed on the Canary Islands. In the US, the Oregon State University and the Northwest National Marine Renewable Energy Centre deployed the Wave Energy Technology-New Zealand (WET-NZ), a 1: 2 scale wave energy conversion device, during August 2012. This shows the diversity of concepts (in-stream or oscillating water column) in the field.

Ocean Thermal Energy Conversion (OTEC) technology relies on a temperature difference of at least 20°C between the warm surface water and colder deep water. This means that only tropical waters have the right conditions for the deployment of OTEC. This technology has the advantage of producing renewable energy on a continuous (non-intermittent) basis. Implementing OTEC demands systems-engineering competencies and industrial capacities that limit the number of players that can be involved in its development and DCNS (France) and Lockheed Martin (US) are the leading industrial players. After completing feasibility studies in La Réunion and Tahiti, DCNS signed export agreements and set up an onshore prototype on La Réunion in 2012. DCNS is also working in partnership with the regional authority of La Martinique and STX France on a 10 MW pilot, which is nearing completion.

Osmotic energy technology uses the energy available from the difference in salt concentrations between seawater and freshwater. Such resources are found in large river estuaries and fjords. The system uses a semi-permeable membrane that allows the salt concentrations to equalise, thus increasing pressure in the seawater compartment. The technology is still in early research and development stages. Statkraftin Norway is one of the few industrial players in this sector. The key to further development lies in optimising membrane characteristics. Today, the membranes generate only a few watts per square metre. The small number of players working on this technology and the need to improve membrane performance and reduce costs therefore point to development prospects in the longer term. Research and development on osmotic power is also being carried out at the Tokyo Institute and in the Netherlands.

Global Efforts towards Ocean Energy as an Emerging Industry

As far as ocean energy is concerned, as of 2012, 530 MW has been deployed while the ocean energy technology initiative of IEA pegs a worldwide potential of upto 337 GW of wave and tidal energy capacity which can be achieved by 2050, along with a contribution from ocean thermal energy conversion. For example, the European Ocean Energy Association forecasts 188 GW of production by 2050 which will fulfil 15 to 20 per cent of the electricity demand of the EU member states'. The UK has set up the Marine Energy Accelerator of the Carbon Trust, to focus on the UK marine energy sector's ability to develop and deploy wave and tidal energy devices on a commercial scale. In June 2011, the UK Department of Energy and Climate Change (DECC) announced that it was investing up to £ 20 million in wave and tidal power to help develop marine energy technologies to support the Marine Energy Array Demonstrator (MEAD) Scheme. The DECC also supports these developments through feed-in tariffs.

According to OES, the deployment forecast could create about 1.2 million direct jobs by 2050. For instance, tidal energy could potentially create 10 to 12 direct and indirect jobs per MW installed, and wave energy could potentially create about 8 to 9 direct and indirect jobs per MW installed. Regarding recent developments and demonstrators actually being tested, the tidal, wave and OTEC are promising and have the potential to be commercialised in the coming years.

Regarding technical barriers to the grid connection, the lack of installation vessels is a barrier to the development of ocean energy which is most frequently cited by stakeholders. A large-scale deployment of ocean energy will thus depend on the sector's ability to address a number of technological and economic challenges.

During the last couple of years, several countries have launched various initiatives aimed at developing ocean energy: a new Danish strategy for the development of wave energy was initiated in 2011 and Japan established its Ocean Energy Technological Development Research Center, which aims to promote ocean energy under the Renewable Energy Plan 2011–2020. In 2012, the French Government presented a roadmap for the development of tidal energy. Canada is also investing in the sector, especially the Nova Scotia region, which put in place a demonstrator site for tidal energy in the Bay of Fundy and released its Marine Renewable Energy Strategy in 2012. The US, China and Korea have also private actors who are investing in the marine energy technologies.

Ocean Energy Resource Assessment: Tidal, Wave and OTEC

Tidal In-Stream Energy (TISE) resource assessment entails the characterisation of the velocity (speed and direction) of the water in the site or area of interest. Normally, a well-calibrated simulation of the tidal flows of the site is the initial step for the TISE resource assessment. This initial resource assessment will narrow down sites which have high energy density and favourable conditions.

Tidal currents are periodic in nature and are caused by the Earth-sun-moon gravitational interaction. The kinetic energy available in the flow of water is proportional to the cube of the velocity of the flow. The validation of the TISE resource will require the use of Acoustic Doppler Current Profilers (ADCPs), which comes in two steps:

(a) Transect Survey Measurements, and
(b) Long-Term Fixed Position Current Profiling.

Step 1 under the TISE Validation employs standard boat-mounted ADCP transect survey measurement methodology to characterise the current through a slice or transect of a channel. Each transect run gives a snapshot of the currents in that particular time period. The long-term fixed position current profiling is studied using the Weibull Analysis and presented as a magnitude histogram and direction as shown in Fig. 2. From such results, the optimum velocity and direction is observed, which is useful to match with the best tidal turbine characteristics in terms of cut-in and rated speed.

Wave power per unit crest denoted as 'W' depends on the wave height 'H_s' and the time period 'T_s' as follows:

$$W = 0.49 H_s^2 T_e \ [W/m] \tag{1}$$

By deploying the wave energy harvester in the field for a one-year period, as shown in Fig.3, the wave amplitude and the wave time period can be monitored and the dominant amplitude, time period and direction can be deduced. By concurrently monitoring the wind conditions, the wind and wave interactions can be ascertained.

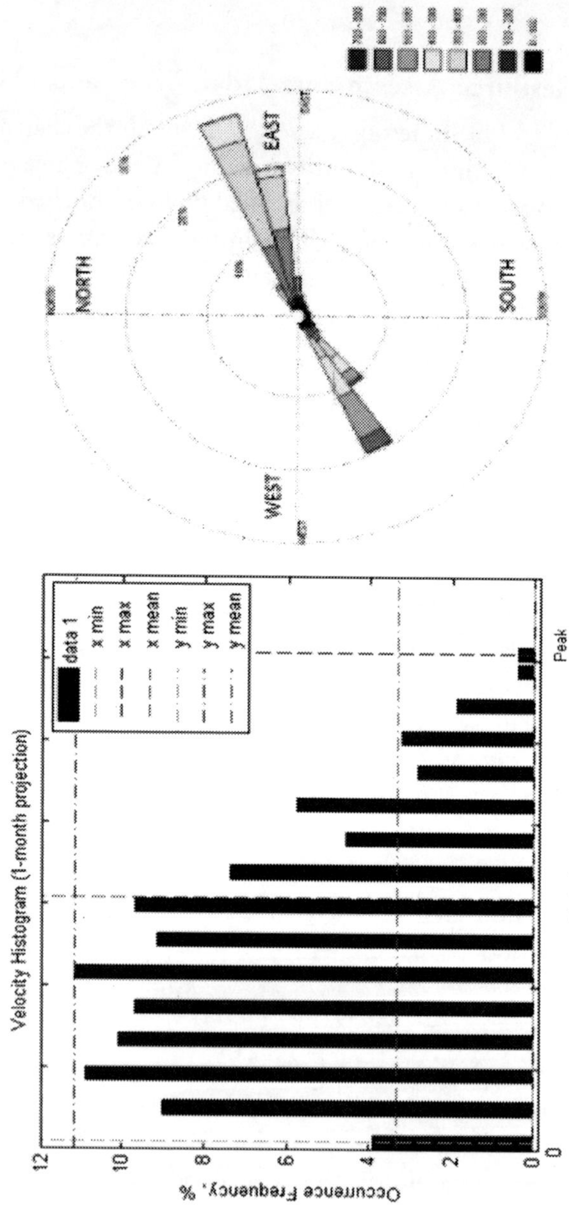

Figure 2. Typical tidal energy resource inference: magnitude and direction

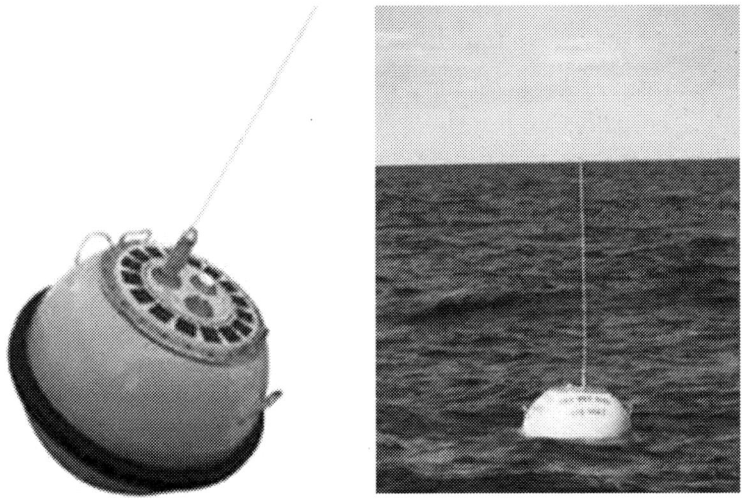

Figure 3. Wave energy monitoring sensors

Site Assessment and Ocean Energy System Sizing

In order to optimise the maximum power output, the ocean energy system should be deployed in the correct location. In the case of tidal turbine, based on the field measurements of the coastal and seabed bathymetry, and using the flow measurements in terms of magnitude and direction, overall flow studies can be performed to identify the high energy potential sites. Fig. 4 shows the CFD predictions of tidal energy region based on field measurements off the Philippines islands.

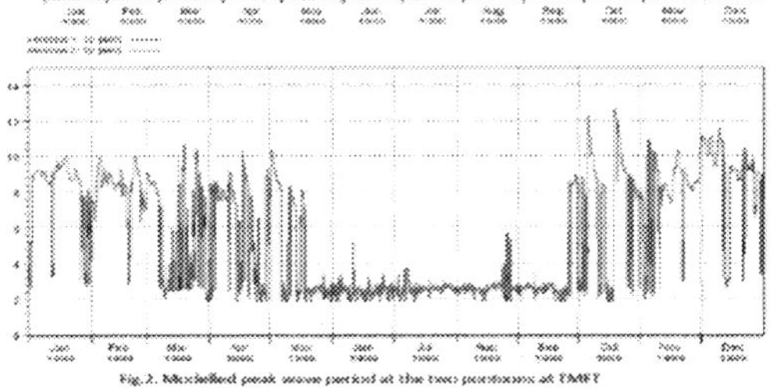

Figure 4. Typical wave energy amplitude and time period results from field measurements

Barriers to Ocean Renewables' Adoption

Today, ocean renewable adoption faces a few key barriers towards deployment. One of the key concerns is its impact on the marine ecology and the possible interference with other marine industries' ocean usage. The lack of sub-sea grids and high cost of sub-sea cables make deep sea ocean energy harvesting impractical and costly compared to fossil fuels. In addition, the lack of a large scale energy storage technology makes it difficult to bring the energy to the shore. Accordingly, researchers are focusing to develop tidal and wave energy-based systems to support remote islands using a distributed smart grid technology coupled with energy storage technology to achieve a high power reliability and power quality.

Among the various ocean energy forms, tidal energy exhibits high predictability, based on the fact that the timing of high and low water levels and the tidal range are routinely and accurately predicted for the use of seafarers worldwide. Technology developers have been investing and developing tidal power-generating devices mainly in Europe and North America. Technology developers and utilities have published plans for specific projects that would amount to a total capacity of 1.7 GW in the tidal stream and 900 MW in wave energy converters. However, the actual amount deployed by 2020 will be far lower, reflecting risks relating to device performance, survivability and cost.

Impacts

Ocean energy systems can potentially have both positive and negative impacts on other marine users such as shipping, fisheries, pleasure boating, scuba diving, sensitive installations, etc. These perceived risks may range from collision, cable entanglement, electromagnetic interference, scouring, ambient noise, etc. Nonetheless, studies are still scarce and the potential impacts are generally presumed. For example, with increasing marine users, the environmental noise pattern has been increasing with shipping, land reclamation, seabed drilling and piling, and hence it is of interest to understand the impact of OES deployment on other maritime users and to marine ecology. For example, the underwater noise caused by the OES during installation and operation may distort the marine species' communication sound spectrum and their ecological behaviour.

A feasibility study is critical to the planning and future development of a marine energy harvester. It allows all interested parties to have an understanding of the key issues leading to a sustainable marine renewable industry. Accordingly, the Environmental Impact Assessment (EIA) focuses on the assessment of the following partial potential impacts:

(a) Hydraulic impacts (changes in current conditions arising from the installation of the facilities at each test site, including the resulting effect on nearby marine facilities)
(b) Morphological impacts
(c) Water quality impacts (changes in flushing, etc.)
(d) Impacts on marine ecology, including marine habitats (e.g. corals, seagrass, mangroves, etc.) and marine fauna (fish, cetaceans, marine reptiles, etc.)
(e) Impacts to aquaculture facilities
(f) Navigation impacts
(g) Recreational and aesthetic impacts
(h) Positive impacts related to renewable energy research
(j) Direct impacts, including loss of invertebrate and soft bottom habitat and
(k) Cross-border impacts

Hydrodynamic Impacts

The deployment of OES close to the coastal line may be suspected for hindrence to the flow of tides and tidal currents and could perhaps create effects that are felt through the region. These depend on the size of the tidal lagoon, its location and its method of operation. The changes in the tidal hydrodynamics may cause changes in the movement of sediments and may impact the natural ecosystem that is presently well adapted to existing conditions. The periodicity of the 'Ebb' and 'Flood' tides for example can set up a near resonant state with the system and may change the tidal amplitudes to a more wider region. Hence, it is suggested that numercial models can be employed to estimate the hydrodynamic impacts. The output of the study helps understand the discharges, water levels and potential towards flood risk. The discharge through each sluice Q_s is given by:

$$Q_S = C_S A\sqrt{2g\Delta H} \quad (2)$$

where C_s is the discharge coefficient for a sluice;
A is the cross-sectional area;
g is the acceleration due to gravity (9.806 m/s^2); and
Δ is the difference in water level.

Typically, efficiency is assumed to be 90 per cent for outflow and 80 per cent on inflow. Using the turbine characteristics as shown in Fig. 5, power output can be ascertained. The largest uncertainties lie in the various inputs assumed during hydromorphological modelling and the best approach is to quantify from the field measurement and input the parameters into the model in terms of flow velocity, turbidity, sea bed soil type and morphology, bathymetry, etc. Such calibrated models help predict sediment transport, potential flooding and other forecasts.

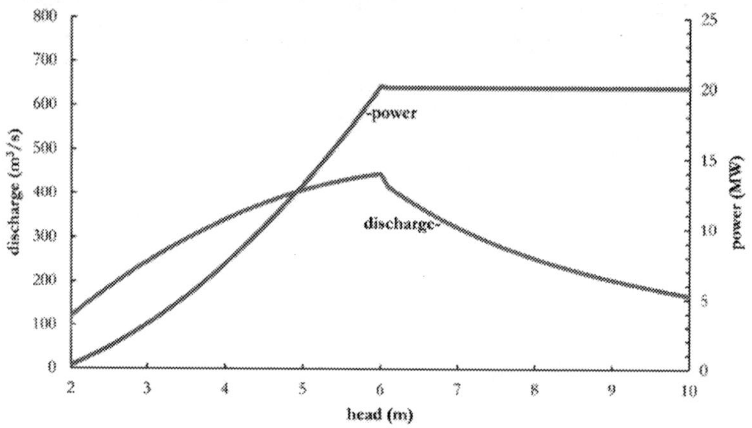

Figure 5. Relationship between head, discharge and power output of a low-head turbine

Coastal geomorphology is important in EIA for two main reasons:

(a) Geomorphological formations and processes are integral components of coastal ecosystems; and

(b) It has direct relevance to problems such as erosion and flooding.

It involves interactions between many components, but the principal factors are waves, tidal regimes and currents. Waves are a major erosive force, and an important ecological factor, especially in the littoral zone. Most waves are wind-generated, and vary in size and force largely in relation to wind velocity, duration and fetch (distance of open water). Wave action is therefore greatest on the coasts exposed to strong on-shore winds over extensive areas of the sea. The wave effects are modified by tidal regimes. An additional problem is posed by rising sea levels associated with climatic warming. The mean global sea level has risen between 10 cm and 20 cm over the last century, and the rise is accelerating with a predicted further rise of up to 6.5 cm by 2100.

Sediment Impacts

The implementation of OES such as tidal turbine and wave energy converters can lead to areas of reduced tidal currents due to energy extraction and also accelerated tidal flows in other areas due to the blockage effects. If the seabed has loose soil, it might lead to changes in the sediment transport regime and impact the seabed morphology in the sandy region. For example, vortices and eddies induced from the offshore turbine tower and foundation structures have known to result in scouring. Studies have shown that energy extraction through array deployment results in a significant impact on the morphology of local sandbanks, but a careful siting of the array may mitigate this impact. However, upto 10–50 MW array peak capacity has less impact on sediment transport. These are case-specific in terms of hydrodynamic, bathymetric and sedimentary properties of sites where the energy extraction is planned.

It is important to characterise the seabed of the locations for the Tidal Instream Device (TISE) deployment in a given site. Not only will the information about the seabed be useful for the device deployment and installation, but also for the support facilities, such as anchoring of mooring lines, placement of gravity foundations, and armouring of subsea cables, etc. The geotechnical data garnered from marine soil investigation can be used for a preliminary design of the foundation and/or support structures needed by various tidal devices and support facilities (e.g. buoy, light markers and control barge). One example of seabed soil testing is to carry out the Cone Penetration Testing of upto 10 metre depth. A laboratory testing of the recovered samples helps ascertain the engineering properties of the sub-surface soil strata encountered and are used for benthic analysis.

Underwater Acoustic Noise

Manmade underwater sound has been known to create disturbances for marine mammals and creatures in terms of sound that exhibits wide bandwith and high amplitude. Examples include the reclamation and piling works, offshore oil and gas industries, sonar-based defence systems, fishing industries' detonations, etc. that have been blamed for noise pollution in the shallow and deep ocean by marine biologists and environmental protection groups. Similar suspision exists during the deployment of a OES as it may hinder the flow of tides and tidal currents near the coasts and lagoons, and also create a sound spectrum that may coincide with marine mammals and organisms.

Table 2. Acoustic Properties of Some Anthropogenic Sounds

Sound	Source level (dB re 1μPa-m)	Bandwidth (Hz)	Directionality
Pile Driving	228 Peak	20-20000	Omnidirectional
Dredging	168-186 Peak	30-2000	Omnidrectional
Small Boats	160-180 RMS	20-10000	Omnidirectional
Large Vessels	180-190 RMS	6-30000	Omnidirectional
Wind Turbine	142 RMS	16-20000	Omnidirectional
Tidal and Wave Energy Devices	165-175 RMS	10-50000	Omnidirectional

Sound is key for marine organisms, turtles, inverteberates and mammals to communicate, determine location, prey identification and predation. There are many different sources of sound from the ocean energy systems. Anthropogenic sounds emitted into the marine environment can potentially affect marine organisms by masking biologically relevant signals (Fig. 6); it can lead to a variety of behavioural reactions. Hearing organs can be affected in the form of hearing loss, and at very high levels, sound can injure or even kill marine life.

Studies investigating sound-induced effects on fish are relatively scarce compared with those on marine mammals, and the results are variable. Adult fish show behavioural responses to the sound waves from seismic activity. Based on the few existing studies, that showed a reduction in catch rates during noise exposure, behavioural response is indicated within a radius of several kilometres from the sound source. If the fish that are on their way to the spawning grounds are exposed to this type of noise, or if they are exposed to the noise during the actual spawning, the effects can have an impact on the fish's spawning success and thereby the recruitment. Avoidance responses of sea turtles to low frequency sounds have also been observed.

As shown in Fig. 7, the tonal frequencies from a ocean energy device such as a tidal turbine can range from 1 Hz to 20,000 Hz which can be deduced indepedently in a lab condition and compared against the audible spectrum of marine mammals and organisms. The sound intensity can be determined for transmission loss as a spherical sound source, and judged for its intensity for any nearby sensitive establishment: for example, the study of interference of a tidal farm on a nearby fish hatchery. Thus the ocean energy device types should be chosen with specific consideration towards the chosen site's marine mammals and other invertebrates' auditorial sensitivity.

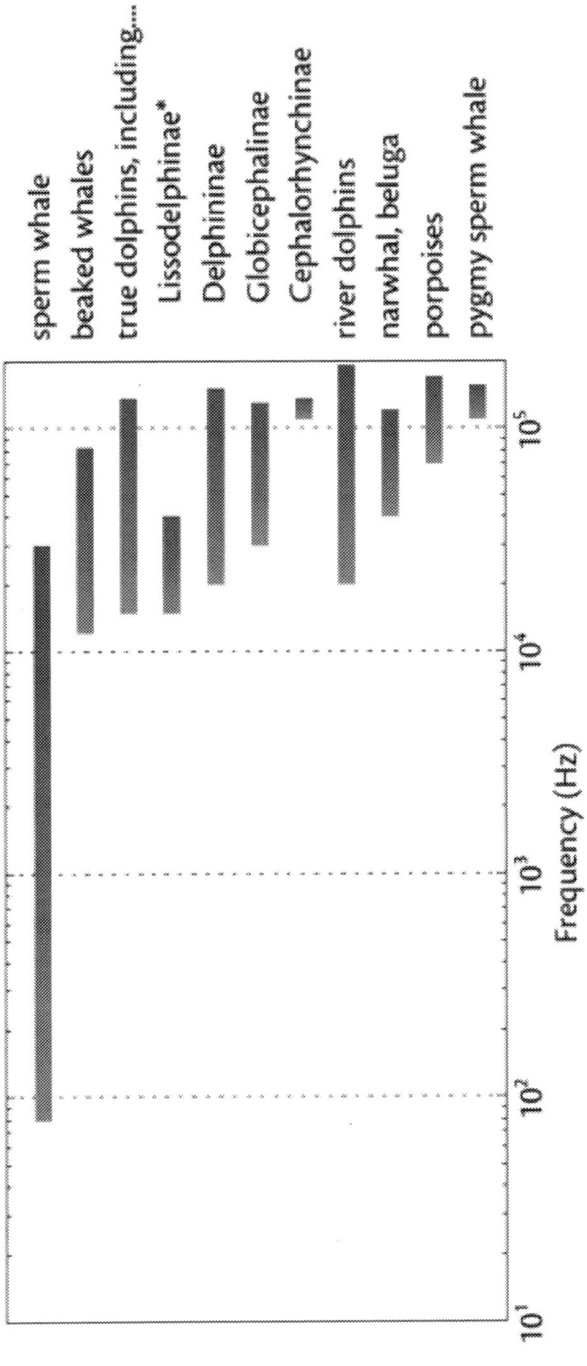

Figure 6. Frequency spectrum of marine mammals

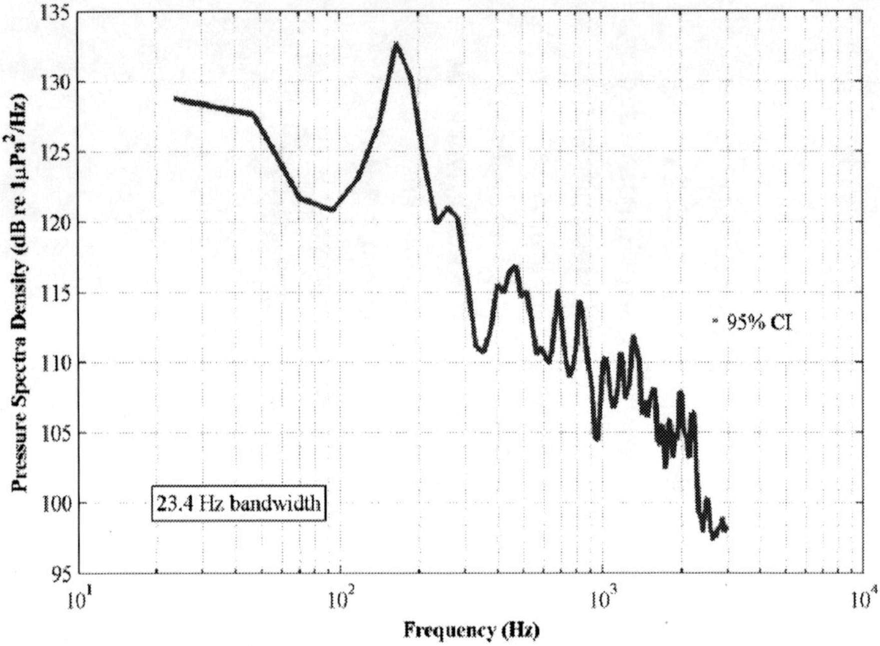

Figure 7. Frequency spectrum of a commercial tidal turbine

Water Quality Impact

A water quality survey should be undertaken, with samples from few stations per location (for example, one at the centre and one at the north, east, south and west boundaries of each test site location) during both spring ebb and flood tides (i.e. 50 samples minimum). All laboratory analyses are to be performed by an accredited laboratory to quantify the following parameters: Heavy metals (As, Cd, Cr, Cu, Hg, Ni, Pb, Zn); Nutrients (NO_3, NH_4, PO_4, Total N, Total P); Oil and Grease; Faecal Coliforms (Total FC, and enterococci); Chlorophyll-a; Total Suspended Solids (TSS); Biochemical Oxygen Demand (BOD); In-situ depth profiles (turbidity, temperature, salinity, dissolved oxygen, pH) and Secchi depth. Such monitoring helps understand the possible contamination due to oil leaks and other pollutants from various coatings, paints and debris that originate from the deployed systems.

Benthic Communities

Most EIAs are likely to focus on the benthic communities of inshore waters.

The environment here is less widely fluctuating than that of the littoral zone, and the seabed is dominated by soft sediments – rocky substrates being normally restricted to narrow zones adjacent to coastlines, and isolated features such as reefs. There is still a range of substrates that largely control the benthic communities, and to a lesser degree those in the water above (which include organisms that depend on the seafloor for food, shelter or reproduction).

(a) There is a considerable variation in water movement, including turbulence, currents and tidal movement.
(b) Near the mouths of rivers and estuaries, salinity may be reduced by freshwater, and turbidity increased by suspended sediments, especially in wet weather.
(c) Turbidity tends to be high also in areas with a muddy or sandy seabed, especially when the sediments are disturbed during storms.

A key stage of risk assessment is therefore quantifying the extent to which the species foraging distributions overlap with the the potential locations of devices within the tidal pass habitats favoured for installations[1] (hereafter referred to as spatial overlap). Tidal passes can usually be divided into either turbine or non-turbine microhabitats based upon the assumption that installations require mean spring current speeds greater than 2 m/s^2.

Seabird Interaction

In some countries, there is a legal responsibility to assess and mitigate impacts of tidal stream turbines on seabird populations (example, The European Birds Directive; 2009/147/EC). A major concern is that deep-diving seabirds (auks Alcidae and cormorants Phalacrocoracidae), known to forage throughout the water column, will collide with the moving components of ocean energy devices. However, collisions between deep-diving seabirds and the moving components of devices will not occur unless these species forage near installations. A simple measure of spatial overlap can therefore be obtained by quantifying and statistically comparing the species' relative use of turbine and non-turbine microhabitats. Common methods include boat-based and aerial-based surveys which could provide accurate measurements specific to a site.

Ocean Energy Systems: Towards Remote Coastal Locations

Energy is crucial to support the essentials of life such as food, water, light, shelter, etc. One of the key challenges of renewables is the need for energy storage to store energy when generation is unavailable. This requires large scale batteries that are of grid level which are not yet fully developed. Hence to support the needs of the remote coastal people, another alternative is to integrate renewables to desalination plants for storing freshwater. Globally, water withdrawals amount to 4,000 billion per cubic metre per year and in some regions, desalination is the main source of water such as in the Middle East and northern Africa.

Desalinated water can be made from two broad types of technologies, i.e. thermal desalination that uses heat to vaporise fresh water, such by integrating with concentrated solar power, and the second technology is membrane desalination, which uses reverse osmosis by pressurising brackish/ seawater through a membrane using electrically powered pumps to separate freshwater using ocean energy-powered systems. As grid level electricity storage is still a challenge, combining power generation and water desalination can also be a cost-effective option for electricity storage when the generation exceeds demand. Typically, a RO plant requires 3.5 to 5 KWh of electricity per cubic metre. Today, desalinated water amounts to 65.2 million per cubic metre per day equivalent to 0.6 per cent of the global water supply. For example, to prepare desalinated water, pilot studies have been successful in integrating desalination systems with renewables such as solar, wind and now with ocean energy systems, such as OTEC and tidal energy.

Using the energy resource potential such as tidal energy, the magnitude and directional data, the rightsizing of the turbine can be achieved through the methodology explained in Fig. 8 to suit a specific site condition. The methodology of sizing an ocean energy system towards a remote coastal location and the EIA constraints are explained in Fig. 9.

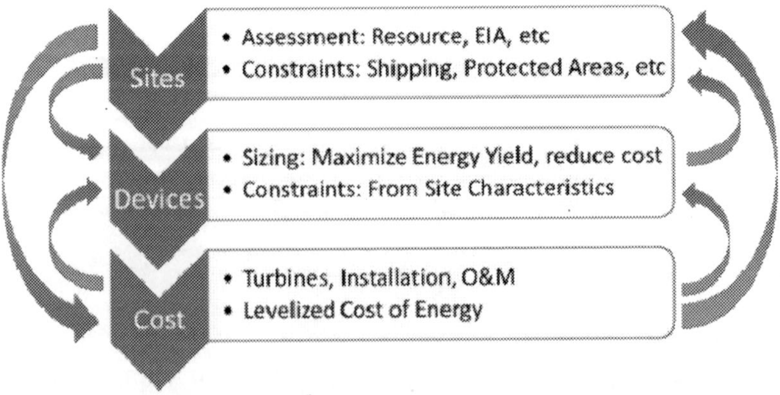

Figure 8. Site-device matching philosophy

Figure 9. Methodology of sizing ocean energy system towards a remote coastal location

The localisation of the ocean energy system should aim to address the specific challenges of deployment and maximise the use of local capabilities. For example, there were specific challenges such as restriction towards piling on the seabed in deploying a turbine in a remote Indonesian coastal site. Hence, an alternative design of mounting the turbine on a barge with self-erecting and deploying capability was adopted such that it provides easy maintenance against biofouling and seaweed entanglement. The barge was supported through catenary moorings. Fig. 10 shows the traditional seabed-mounted turbine and the barge-mounted tidal turbine design deployed in Indonesia to minimise deployment and operational costs to suit the remote coastal needs.

Regional Supply Chain Creation

One of the key barriers to ocean energy adoption is the lack of a supply chain to promote the ocean energy harvesters towards mainstream electricity generation. Learning from the case of wind and solar energy, indigenisation has to be promoted in a given region. Countries such as India have a significant know-how in other technologically similar industries such as shipping, naval defence, machine tools and offshore oil and gas industry. India has significant skilled people specialised in naval architecture, advanced composite materials, precision metallic component fabrication, offshore structures, control systems and other key domains that are crucial to indigenise the OES development. Hence, a test bedding of global designs can be a good start with the globally available system design and a subsequent follow-up with an indigenous design development to suit to the local sea and weather conditions. In order to reduce the costs further, the materials and technologies should be locally sourced such that the developmental progress is not hindered and expensive due to imports. Since knowledge is generally classified as explicit and tacit in nature, research shows that if a new industry is collocated with various technologically

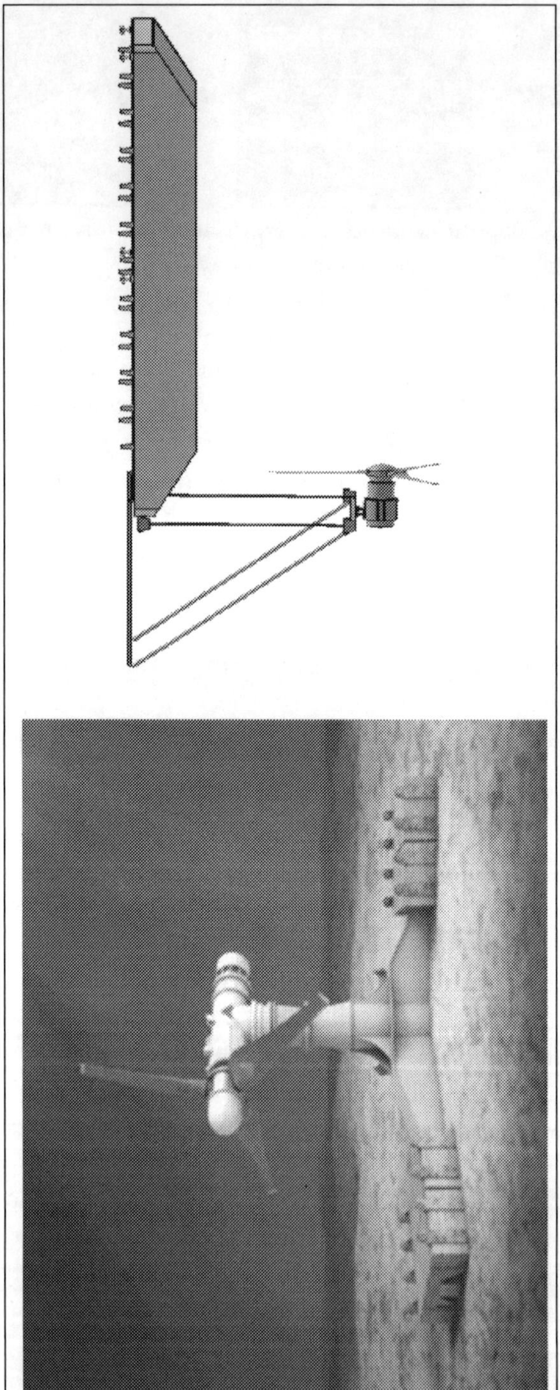

Figure 10. (a) Traditional seabed-mounted turbine; (b) Proposed barge-mounted tidal turbine

related firms as an industrial cluster, it promotes knowledge spillover through skilled people on the job and migration from different related industries, thereby promoting domestication and indigenisation that results in technology scaling and cost reduction. Today, the success of India's solar and wind energy is evidence towards such indigenisation of renewables' production and adoption.

To induce the new ocean energy industry evolution, the government should:

(a) Initiate domestic market creation through subsidies to induce large ocean energy-based power plant adoption
(b) Develop a domestic supply chain through R&D subsidy policies that could promote localisation and
(c) Promote a close working relation between the industry, market and academia (such as with IITs and NIOT-type institutions) to promote technology customisation towards regional energy needs

Such efforts create jobs, meet local energy needs and ensure a sustained regional growth of the ocean energy industry.

Conclusion

The ocean energy is a free and a clean source that could emerge as a promising alternative energy source for the region. Tidal, wave and ocean thermal are promising energy sources when their resource availability is quantified along with water depth, seabed and coastal bathymetric conditions, wave and tidal load conditions, biofouling and corrosion challenges. Accordingly, devices developed in temperate climate need to be tuned towards tropical needs, which could be undertaken by local firms using similar technology from similar industries. This will help achieve localisation and supply chain creation and support the ocean energy adoption.

Remote coastal and island regions which are presently supported by fossil fuels can be early adopters of the ocean energy systems. The energy harnessed can significantly contribute to the energy and water independence and can provide the necessary resilience to the region. To ensure a good healthy marine ecology and minimal hindrance to the other marine users, a responsible environmental impact assessment is recommended. This would also need baseline information and continued real time monitoring of the ocean health. It would also require necessary approvals from the government agencies. The resource identification and EIA help in identifying various constraints and to ascertain the optimum device design and size to minimize the asset cost, as well as the operational cost for the maximum annual energy yield.

REFERENCES AND NOTES

1. Huynh, B.T., T. Tjahjowidodo, Z.W. Zhong, Y. Wang and N. Srikanth, 'Nonlinearly Enhanced Vortex Induced Vibrations for Energy Harvesting', *Advanced Intelligent Mechatronics (AIM)*, 2015 IEEE International Conference, 2015, pp. 91-96.
2. Quirapas, M.A.J., N. Srikanth, M. Abundo and H. Lin, 'Electricity from the Ocean: A Path to secure Energy in Southeast Asia', 5th International Conference on Sustainable Conference on Sustainable Energy and Environment (SEE 2014): Science, Technology, Innovation for ASEAN Green Growth 19-21, ASEAN Green Growth 19-21, Bangkok, Thailand, 2014.

4

Marine Leisure Industry and Infrastructure

Sujay Chohan

India has an incredible opportunity to develop the boating industry in an environmentally sensitive, planned and regulated manner in line with the tenets of the Blue Ocean Economy. India is a country with vast coastlines, inland waterways, spectacular island groups, suitable climatic conditions, a growing economy and a large young population. India has the untapped potential to be home to the next French Riviera and establish itself as a primary leisure boating destination for both local and global tourism. The development of water sports on the beaches and lakes offers low-cost recreation opportunities for the masses while encouraging micro-entrepreneurs, as witnessed in the spurt of water sports activities in the recent years in Goa and along the coast of Maharashtra. Internationally, boating is a large contributor to the local economy. If the government recognises boating as a sunrise industry, develops a comprehensive national boating policy and invests in the development of marine infrastructure, India will greatly benefit in terms of job creation and significant contribution to the GDP. Nautical tourism is an integral part of the tourism industry and has numerous indirect effects that are more important than the direct, thus supporting the development of many other industries. To kick-start the growth cycle, the first step is to create a national boating policy to regulate the industry in terms of safety, security, marina development, cruising, charters, manufacturing, crewing, rescue and the rationalisation of import duties and local taxes.

Introduction

Leisure boating in India has the potential to create thousands of new jobs and become a multi-billion-dollar industry over the next decade. There are several reasons for this view. First, India is the fastest growing economy in the world and over the next decade, the average age of Indians is likely to be around 29 years,[1] all of whom are looking for jobs and new avenues of recreation. Second, India has arguably the best weather conditions for boating with an average of eight months' sunshine compared to only four months in the Western world. Third, India has an untapped coastline of over 7,600 kms, tens of thousands of kilometres of internal waterways, lakes, rivers and backwaters; besides, the pristine and spectacular island chains of the Andaman & Nicobar on the east coast and the virgin atolls and reefs of Lakshadweep on the west coast are the most suitable for recreation. In short, India is probably the only country on the planet with such statistics making it arguably the last large leisure boating and nautical tourism market in the world.

India has a glorious past as a seafaring nation and has a tradition of building some of the famous ships for the British fleet. However, it was only in the inaugural "Mumbai International Boat Show" held in February 2007, that leisure boating took off spectacularly. By the third show in 2009, over 300 large and small yachts were sold, which were valued at over US$ 100 million. A latent demand appears to have been tapped and this led to the creation of a new breed of entrepreneurs in the form of boat dealers, charter operators, technical and maintenance service providers, as well as water sports operators offering new recreation avenues to local tourists.

Instead of recognising boating as an opportunity to create new greenfield jobs and working with industry bodies, like the Yachting Association of India, to encourage and promote leisure boating and nautical tourism, the government chose to ignore it by branding it as elitist and doubled the duties on goods imported for recreational boating in 2013, effectively stalling the growth of the industry. The leisure boating industry cannot succeed without government support. To grow, it needs marinas and infrastructure in terms of jetties and access pontoons. Further, a national leisure boating policy covering various aspects of the industry from rationalising the import duties of products not manufactured in India and incentives by way of tax holidays is required to encourage the boat building and ancillary industry.

Local manufacturing is more expensive, despite having low-cost labour in India because of the high excise, sales tax, and VAT being levied on locally made boats, over and above the import duty paid for the raw material and components. Compare this with the growth of the industry in China, Thailand and Sri Lanka, all of whom have created thriving boating industries in the last decade by recognising the potential and supporting it with marine infrastructure and tax incentives. China, for example, has comprehensively overtaken India after a slow start by encouraging marina-led residential projects resulting in over 10,000 marina berths[2] created over the past decade, as compared to only 43 berths in India. India has only one marina in Kochi, the Kochi International Marina, and Mumbai, being a global island city, has none. Marinas are important because they significantly bring down the Total Cost of Ownership (TCO) of boats and yachts, ensure security and tracking of boats and drive tourism. For example, a significant percentage of the GDP of the South of France comes from leisure boating with marinas dotting the coastline.

The present government is the first federal government to have recognised the opportunity offered by the un-utilised water resources for the transport of both goods and people and has charted an ambitious path to achieve this. Although the government has moved aggressively to develop these facilities, leisure boating and nautical tourism find no mention in the policies. There is tremendous opportunity for India to promote leisure boating and nautical tourism and the Indian government needs to draft a national boating policy. It can borrow a page from the hugely successful international policies while building in environment safeguards and the need to provide the much needed clarity for investors and entrepreneurs alike. The national policy for boating should cover the development of waterfront, marina infrastructure, a uniform registration process, guidelines for boat building, policy to govern commercial charter operations, free movement of super-yachts, environment safe guards, skill development, etc.

The policy should include single-window clearance for marina development and relaxation in the Coastal Regulation Zone (CRZ) guidelines to facilitate the development of real-estate led marinas and super-yacht harbours. Agencies like the Yachting Association of India (YAI) under the chairmanship of the Indian Navy is the apex body of yachting in India and could be considered to lead this initiative and table a comprehensive policy on boating. Organisations like the National Maritime Foundation could also help research and influence policy.

An Overview of India

Economy

India is the seventh-largest country, having a trillion-dollar economy status and an overview of India's economy is presented in Table 1. India's Gross Domestic Product (GDP) growth rate was 7.6 per cent[3] in 2015–16 and 7.1 per cent[4] in 2016–17.

Table 1. Overview of India's Economy

	2007-08 (US$)	Current (US$)
GDP at Factor Cost (Current Prices)	1.06 trillion	2.30 trillion[1]
Forex Reserves	309.7 billion	355.5 billion[2]
Imports	235.9 billion	379.6 billion[3]
Exports	155.5 billion	261.1 billion[4]
FDI Inflows	29.9 billion	40 billion[5]

Demography

India is the world's second most populous country with a population of nearly 1.3 billion and a population growth rate of 1.2 per cent p.a.[10] By 2020, India will have 900 million people of working age, and the average age of its citizens will fall to 29 years.[11]

Income Distribution

Measured in domestic currency, wealth has grown rapidly in India since 2000, except during the global financial crisis, at an average of 8 per cent. Wealth per adult was US$ 4,350 in mid-2015. Only 3 per cent of India's adult population falls into the middle-class wealth band, but its middle-class wealth has grown by 150 per cent since 2000 to US$ 780 billion, representing 23 per cent of the country's wealth. The average wealth of the middle class is ten times the level of those in the rest of the population.[12]

India has over 2,00,000 High Net-Worth Individuals (HNIs)[13] with a large young population with the propensity to spend on luxury. As of 2015, India had 2,083 ultra-high-net-worth individuals with more than US$ 50 million net wealth – 3 per cent higher than 2014.[14]

According to the Asia Pacific 2016 Wealth Report, by New World

Wealth,[15] India was ranked among the top five Asia-Pacific countries in terms of the number of HNIs. India is home to the fourth largest population of millionaires in the Asia-Pacific region, with 2.36 lakh such HNIs. By 2025, India is expected to see a 105 per cent growth in the HNI population to 4,83,800 from 2,36,000.

These globally aware and wealthy individuals were the first to embrace the leisure boating opportunity. However, over the years they have lost interest due to the complete lack of infrastructure, high taxes and harassment by various government agencies in the absence of a comprehensive national boating policy.

Climate

India's weather is perfect for boating through the winter and summer months and the lean season is only during the monsoon period between June and September. This compliments the European boating season, which closes for the winter in October, ensuring that the boating enthusiasts can travel to India during the colder months. Presently almost all super-yachts and sailors skip India, as there is no infrastructure to support these vessels, thus losing millions of dollars of annual revenue and opportunities to create new jobs.

Coastline and Inland Waterways

India has 7,516 kilometres of coastline comprising of the peninsula, the Andaman & Nicobar and the Lakshadweep islands, i.e. nearly 14,500 kilometres of inland navigable waterways, 12 major rivers and approximately 140 small and big lakes. The west coast of India from Gujarat to Kerala has some of the most beautiful naturally sheltered harbours and creeks having the potential to emerge as the primary boating destination in the world. With beautiful virgin beaches and blue waters, the potential for sailing and water sports is enormous. The coastline from Mumbai to Goa can be developed along the lines of the French Riviera, significantly enhancing the GDP of the region and establishing it as a global and regional recreational tourism hub. Similarly, the inland waterways and backwaters of Kerala have a huge potential for being turned into hubs for nautical tourism.

Tourism

Table 2 shows the growth of tourists in India.

Table 2. Tourism in India

	2007	2015[16]
Foreign Tourist Arrivals	5.08 million	23.33 million
Domestic Tourist Visits to all States/UTs	526.57 million	1431.97 million
Foreign Exchange Earnings from Tourism	US$ 10.73 billion	US$ 21.45 billion
Share of India in International Tourism Arrivals	0.56 per cent	0.68 per cent

The federal government recognises the need to further enhance the footprint of the tourism industry. Boating and water-based tourism should become an integral part of this policy. The introduction of organised cruising, game fishing, scuba diving and other recreational boating activities would also open the market up to many more opportunities. However, the government also needs to be sensitive to protecting the environment, and needs to ensure that only high-end tourism is allowed especially in the island groups of the Andaman & Nicobar and Lakshadweep to ensure the protection of the spectacular natural beauty and coral reefs in these areas.

Yachting Industry, Boating and Water Sports

Global Boating

Internationally, many boating hubs exist where the industry is an integral part of the culture and economy, and a boating lifestyle is ingrained amongst the locals, be it the French Riviera, Gold Coast, the Mediterranean, the Caribbean, the Maldives, Mauritius, Thailand and Dubai. This is supported by favourable laws, safety guidelines, stringent environment laws and regulations, infrastructure and incentives for local manufacturing and servicing.

According to estimates made by Deloitte and Fondazione Altagamma for 2015,[17] the global market of new boats is valued at 19 billion euros, recording an increase by 12 per cent versus 2014 (17 billion euro). 8,00,000 new boats were sold in 2015 as compared to 7,00,000 boats sold in 2014, thus registering a growth by 14 per cent. The 2016 order book of super-yachts (boats longer than 30 metres) is 295 boats with an overall value of 8.7 billion euros, and 44 per cent of these are new orders. The pre-owned boats market has an estimated value of 14 billion euros, while the refit and repairs and super-yacht charter market has an estimated value of 1 billion euros.

In 2015, the global market of new boats was valued at 19 billion euros, an increase by 12 per cent from 2014.[18] It is expected to grow to approximately US$ 25 billion by 2020[19] (Only sail boats and power yachts). Given below are a few snapshots highlighting the tremendous opportunity for the Indian leisure boating and nautical tourism industry.

(a) *USA*: Around 34 per cent of the total population of the US participates in recreational boating on an annual basis. In 2010, about 75 million people in the US participated in recreational boating. The number increased to over 88.5 million by 2013.[20] The US sales of recreational boating products are expected to increase by 6.1 per cent per year to US$ 13.1 billion in 2018.[21]

(b) *Europe*: 36 million boaters, 6 million registered boats, over 48+ million persons enjoy water sports, a turnover of 20 billion euros and 4,500 marinas with 1.75 million berths. The industry directly employees over 2,80,000 people and is largely driven by small and medium businesses.[22]

(c) *China*: Successfully set up a boating industry in the last decade with 4,000 registered yachts, 100 marinas with 10,000 marina berths and with 15,000 more berths being planned.[23]

(d) *Thailand*: In 2004, the import duty on leisure vessels was reduced from 245 per cent to 0 per cent, which resulted in a 350 per cent increase in yacht registrations from 2003 to 2004. Thailand now has a well-developed industry with 4 major marinas, more than 20 yacht charter operators, 30,000 licensed captains and crew members, more than 800 permanently berthed yachts, and a robust manufacturing base.[24]

(e) *India*: Approximately 500 power boats and a handful of yachts, one 43 berthed marina and almost no nautical tourism, but having an immense potential that is waiting to be developed.

Major Destinations

Location is an important factor for a flourishing boating industry. India sits astride international sea lanes with the Indian Ocean lapping at its feet. From a leisure boating perspective, India is the natural first port of call for yachts traversing the Suez and its proximity to Dubai, Maldives, Thailand and Malaysia make it a natural leisure boating and nautical tourism hub. India has a number of regions which can be developed as boating hubs.

(a) *Mumbai*: Mumbai with its natural harbour and surrounding sheltered coastline is the main location for yachting and leisure boats, given that it is the financial capital of the country. The present government has initiated the development of tourism on defunct lighthouses. The Kanhoji Angre Island Light House, 14 nautical miles south of Mumbai, is all set to drive nautical tourism. Water sports centres in and around Mumbai provide experiences such as sailing, para-sailing, water skiing, canoeing, kayaking, jet skiing, jet boating, windsurfing, rowing, etc. A majority of the boats in Mumbai are moored in the open sea off the Gateway of India and are commonly used for harbour cruises or visits to the Mandwa Jetty. However, there is no marina in the city and this has inhibited the growth of the boating industry.

(b) *Goa*: Goa is a tourist-friendly hub with innumerable beaches, making it one of the most popular destinations for domestic and international travellers. It handles 12 per cent of all foreign tourist arrivals in India and has a thriving water sports industry but needs marinas and jetties to promote game fishing, charters and diving activities.

(c) *Mumbai to Goa Coastline*: The numerous creeks along India's west coast lend themselves ideally to set up boat havens and a marina-led real estate. Along with pristine beaches and blue waters, there is every possibility to develop the coast along the lines of the 'French Riviera', driving a huge potential for boating and nautical tourism for local and international visitors.

(d) *Kochi*: It is the second largest city in Kerala with a port located on the busy international sea route just 11 nautical miles from the Gulf-Singapore-Far East route and 70 nautical miles off the Suez-Singapore-Far East route making it the maritime gateway for peninsular India and one of the best ports in Asia. India's first permanent marina, the Kochi International Marina, was built by Ocean Blue on Bolgatty Island in Kochi in 2010, with a 22-room clubhouse, restaurant and a 9-hole golf course. The marina has since attracted hundreds of sailors, as it offers full-service marina facilities and is an ideal stopover for global voyages for exploring India. The marina is due for an upgrade and expansion.

(e) *Lakshadweep:* A mere 200 nautical miles from Kochi lies the Lakshadweep archipelago, located atop the Chagos-Maldives-Laccadive Ridge, a vast submarine mountain range in the Indian

Ocean, which also forms a terrestrial ecoregion, together with the Chagos and the Lakshadweep. Unlike, Maldives, the atolls and coral reefs of the Lakshadweep have remained undisturbed, as there are very few facilities on the atolls. To preserve this pristine ecosystem, only limited high-end eco-tourism should be permitted on self-contained and certified charter vessels i.e. sailing catamarans and live-aboards, allowing guests to enjoy the spectacular coral reefs but ensuring that all waste is carried back to the shore.

(f) *Andaman & Nicobar Islands*: Approximately 750 nautical miles off the east coast of India, the islands are an Indian archipelago in the Bay of Bengal. Roughly 572 islands stretch over 750 km and are known for their palm-lined, white-sand beaches, mangroves and tropical rainforests. Coral reefs supporting marine life such as sharks and rays make for popular diving and snorkeling sites. Indigenous Andaman Islanders inhabit the more remote islands, many of which are off-limits to visitors. Andaman & Nicobar also has the distinction of having the only active volcano in India on the Barren Island.

With its pristine waters and coral reefs teeming with life, the Andaman & Nicobar Islands can be developed as a high-end boating and nautical tourism destination, to encourage diving charters, game fishing (catch and release) and cruising holidays like in the Caribbean. A well-thought-out policy is essential to ensure that the pristine environment is protected. The Andaman & Nicobar also sees the transit of super-yachts every year en route from the Mediterranean and Maldives to Thailand, Malaysia and Singapore, currently bypassing the islands completely, due to lack of facilities. Super-yachts typically contribute hundreds of thousands of dollars to the local economy during stopovers and India is missing out on these revenues. In addition to the above-mentioned locations, there is a huge potential in India's coastal states in inland waterways and backwaters for the development of boating and water sports.

Market Opportunity and Demand Potential

India is arguably a large leisure boating and nautical market in the world. There are various factors that will contribute to this growth:

(i) The mild sunny weather, especially on the west coast of India with conditions conducive to boating for eight months of the year.

(ii) A growing economy promising to create more HNIs and individuals with expendable income.

(iii) A young population with limited recreational avenues, excited to indulge in a new lifestyle.
(iv) India's proximity to international sea routes.
(v) An ability for anyone to experience boating with lower-ticket water sports experience, a mid-ticket yacht charter experience and eventually buying options for high-involvement customers.
(vi) A lack on any nautical tourism options for international tourists who would love to visit various islands and the pristine coastline.

The Market Segments

Recreational Boating (Power Boating, Sailing and Personal Watercrafts)

Presently, there are over 300 boats registered in India, comprising of over 150 speedboats and 120 luxury yachts. Most of these are foreign brands and the popular size ranges from 15 to 35 feet. There are three factors driving the growth in this segment:

(i) Growing Purchasing Power: India continues to be the bright spot in the global economy growing at 7.1 per cent and more young people now have a disposable income.
(ii) Young Population: India is slated to have a booming young population of individuals who have grown up in the internet age, are well informed and are extremely health-conscious.
(iii) Lack of recreational options: Presently there is precious little to do other than dining, movies and malls. With growing global awareness and disposable income, the youth is seeking active lifestyles, which boating offers.

Charter Services

Boating is still a newer experience for Indians and for the uninitiated, owing a yacht would entail large capital investments as well as recurring expenses for the maintenance and crew due to the absence of marinas. The best way to provide a taste of the boating lifestyle at a fraction of the cost is by providing yachts for charter, either for a few hours within the local harbour, or for a few days to travel from one place to another. Charter services are a major part of the boating economy, especially in locations such as the south of France or the Caribbean and in the Maldives, Thailand and Sri Lanka.

Charter business in India was thriving but a short-sighted law enforcement has forced service providers to shut shop citing security reasons. To drive tourism, the government needs to address the concerns of the industry and needs to draft a charter policy laying out rules and regulations including safety and security guidelines. Charters will drive revenue to the exchequer by way of service tax and will introduce first-timers to boating and sailing and providing the much needed alternatives for tourists visiting coastal and inland waterfront cities.

Sailing

Sailing as a sport is growing in India and India has competitively participated at various international sporting events such as the Asian Games, Olympic Games, etc. There are many initiatives by the Yachting Association of India to promote the sport of sailing.

Water Sports

Water sports would typically be the first avenue for an individual to experience boating by providing access to the masses, debunking the myth that it is an activity only for the elite. The water sports industry in India is estimated to be about INR 100 crores. There are currently around 150 small and medium water sports operators in India, out of which only 50 per cent are licensed. Primarily 90 percent of the operators in India are family-run small-time businesses, and operate without any safety, quality or regulatory guidelines.

Marinas and Infrastructure

One of the key reason why boating has not taken off is because of the lack of access to water. There is simply no safe way to access boats in India (and this includes commercial ferry operations). Although the situation is changing, the speed of the work must be enhanced. Alongside the Mumbai International Boat Show in 2007, a floating pontoon system was set up off the Gateway of India with a gangway for embarkation onto boats, to demonstrate the ease of stepping on and off in a safe manner. Similar pontoons were created for events such as the Volvo Ocean Race in Kochi.

India's first marina, with 43 berths, was built in 2009 in Kochi and serves as an ideal stopover point for boats travelling internationally with state-of-the-art facilities in the form of freshwater, electricity, fuelling and sewage pump-outs. The hope was that a domino effect would be created and after the development of the first marina, many others would emerge across the country.

Unfortunately, despite multiple opportunities where the necessity and benefits of marinas and secure jetties were demonstrated, progress has been slow over the last decade and only 43 berths have been created in India compared to over 10,000[25] in the same time frame in China. The government needs to develop a comprehensive policy for marina and water-front development (like it has for ports) and invest in nautical infrastructure to promote the industry. Policies around CRZ clearances, the development of marina-led water front real estate to recover investments in the marinas are some of the key areas which need to be overhauled.

Services

The boating industry would be crippled without trained and technically sound personnel providing servicing and maintenance for the imported and locally manufactured boats and personal watercrafts. Regular maintenance and easy access to servicing will enable more people to invest in own crafts without the fear of the product being mismanaged or deteriorating into disrepair. A trained crew is also provided to the boat owners to ensure safety while on water. There is a need for state-of-the-art service yards with facilities for maintenance, repair and lifting and lowering of boats to ensure that they are safely stored during the monsoon months. As the number of boats in the market increase, there will be a surge in demand for skilled manpower and service facilities to ensure the efficient operation and maintenance of the boats, leading to skill-training opportunities and employment options.

Finance

The boating industry being nascent, banks and financial institutions have yet to get involved. A parallel can be drawn to the Indian auto industry which took off in the mid-90s with the introduction of low-cost 100 per cent finance, which is now the norm for vehicle purchase. In the Western economies, bank financing plays a major role in the growth of the industry from purchases of personal watercrafts (e.g. Sea-Doos) to large super-yachts (e.g. Azimut Benetti) worth hundreds of millions of dollars. Financing gives the buyer the ability to enjoy the lifestyle while paying for it over a period of time and is one of the keys to the eventual growth of the recreational industry by making purchasing a feasible option for a larger cross-section of buyers.

Imports

Currently the industry is dominated by imported boats, be it personal watercrafts or large yachts. In most cases, international yacht manufacturers appoint dealers who have a presence in various states and countries. Customers who buy high-value products are generally well-informed, well-travelled and have an in-depth understanding of the available brands and models. Hence, local dealers are the best way forward since they understand the product specifications and can match the requirements of the customer and the local conditions in terms of weather, water body where the product would be used and the facilities available within that geographical area.

Local Manufacturing

There are a few regional manufacturers that make cheap, inefficient and unsafe boats due to the lack of regulatory and safety guidelines. However, in the past decade, since the Mumbai International Boat Show and the inflow of international yachts in the market, a few corporates have put their weight behind boat building. With the boost given to local manufacturing as part of the "Make in India" initiative, the government also needs to rationalise local taxes and provide incentives to local companies to manufacture world-class products. This is a slow process since there is no eco-system to support boat building in India and almost all raw materials have to be imported including fibre, resins, components and engines. India can take a leaf out of initiatives put in place by Sri Lanka and Thailand where local boat building has flourished. The potential for "Make in India" boats is huge. Based on experiences in international markets, the average understanding is that each boat manufactured locally creates 2 to 5 direct jobs and 10 indirect jobs. Overall, thousands of jobs have been created in various sectors of the industry since the inaugural Mumbai International Boat Show a decade ago.

Regulatory Framework

The regulatory framework for the boating industry is practically non-existent. The industry has at various forums stressed the need for a uniform policy to govern and regulate the industry to help it grow. In 2013, the honourable Finance Minister increased the import duties from an affordable 26 per cent to 49 per cent which affected the struggling industry which was already reeling from the 2009 global economic crisis.

Government Policy

Barring a couple of states having a limited policy for boating and watersports operations, there are no established national guidelines which in turn deters serious stakeholders, making the industry seem like a high-risk investment. Most importantly, safety and operational quality are compromised.

Environment Policy

To ensure India does not make the same mistakes as the West, the Indian policy must strive at responsible boating by embedding environment safeguards as part of the policy, making everybody from manufacturers, service providers, equipment suppliers and users responsible for protecting the environment. Building marinas which have waste-disposal handling and fuelling facilities will go a long way in ensuring environmental protection (at present there are no on-water fuelling stations for leisure boats, making this not only a hazardous activity but also highly polluting.)

Registration

The registration process for boats is quite ambiguous. It is important to define detailed segments under which the yachts, speedboats and personal watercrafts can be registered to ensure that the licenses and certifications are accurate and relevant to the product. Applying shipping laws to pleasure boating is inefficient and can lead to grey areas, resulting in safety and security issues.

Certification

Internationally, there is certification for pleasure boats which defines pertinent limitations such as the type and number of crew necessary on board, the limitation on the number of people on board, the distances that the boat can travel, the locations that a boat can travel to, etc.; however, a disorganised small-time manufacturing set-up where private yards are making one-off boats, results in risks to boats being launched in the market without any proper certification, neglecting on-water safety. The European Union "CE" certification is a widely recognised and respected certification for pleasure and recreational craft. India can adopt similar guidelines for building and certify small boats instead of applying archaic shipping acts meant for ocean-going vessels which make it impossible to comply with.

Imports and Taxation

Custom duties on the import of boats is at 49 per cent, stifling the inflow of yachts and pleasure boats into the country from foreign-made manufacturers even if they already have local dealer representation. In addition, any components required for local manufacturers also attract a high customs duty, excise, octroy duty etc., which exponentially increases the production cost and in turn, price the Indian manufactured boats above the competitively priced alternatives.

Industry Bodies

(a) The Yachting Association of India is affiliated to the International Sailing Federation (ISAF) which is recognised as the governing authority for sailing worldwide by the International Olympic Committee (IOC). The YAI is officially recognised by the Indian Olympic Association and the Ministry of Youth Affairs and Sports, Government of India, as the governing authority for sailing, windsurfing, motorboating, powerboat racing and personal watercrafts, at sea and on inland waterways in India. The YAI Certifications are specific to the type of boat (i.e. keelboat, multihull, etc). The priority of the YAI is safety and it is responsible for certifying the training centres to ensure that standards are maintained in terms of facilities and equipment.

(b) The Indian Olympic Association is the apex sports organisation of the Olympics Sports in the country and is responsible for the Indian contingent's participation in the Olympic Games, Commonwealth Games, Asian Games and the South Asian Games. The sport of sailing has been dominated by the Indian Navy and Army Clubs with the support of a few private yachting clubs.

(c) The Maritime Board and Captain of Ports are governmental organisations set in different states of India. The board is tasked with the administration of ports and harbours, conservancy, licensing of crafts, levying of fees, regulation and control of traffic.

Job Creation and Skill Development

The boating industry encompasses wide market segments which India has begun to explore. As various commercial enterprises begin to flourish, the industry will require skilled labour, and therefore the skilling of personnel for the boating industry has to be included in the boating agenda. Moreover, the boating industry will also require specific training for certain functions, providing a new skill training opportunity. This would be a great avenue to enhance the employability of individuals from the younger generation of family businesses (such as fishermen etc.), who can put their existing knowledge of our waterways to use at a larger scale.

Fig. 1 shows the creation of employment and opportunities for skill development related to the boating industry. It is interesting to note that the industry not only provides direct employment within the various facets of boating, but also intermingles within and outside of the boating universe and supports or enhances various other businesses, including but not limited to real estate, entertainment, tourism, food and beverage, retail, etc.

SWOT Analysis of the Industry

STRENGTHS

- Awareness around boating has grown over the last decade in India,
- Vast coastline
- Extensive network of back waters, rivers, freshwater and saltwater lakes
- Spectacular islands groups for the Andaman and Nicobar and the atolls and reefs of the Lakshadweep Islands
- 8 months of sunshine and fair weather
- High economic growth rate.
- Latent demand – lack of recreational options
- Young population
- Availability of talented labour-pool ready for skill training- High growth rate of HNIs
- Natural creeks on the western coast can be developed into real estate marinas and boat havens

WEAKNESS

- Lack of a national boating policy
- No coherent policy around charters and cruising
- No environment policy on responsible boating
- No uniform registration process
- Absence of marine infrastructure to access the water (jetties and pontoons)
- Lack of marinas
- CRZ laws stall development of lifestyle marinas
- Applying archaic Inland Vessel Act to Boating
- Applying archaic Telegraph Act making it impossible to procure Radars or VHF radios essential for safe navigation
- High customs duties at 49% making imports expensive
- Lack of quality manufacturing
- Lack of clarity on regulatory responsibilities in terms of registration
- Lack of nautical tourism
- Towing of personal water craft and boats not permitted

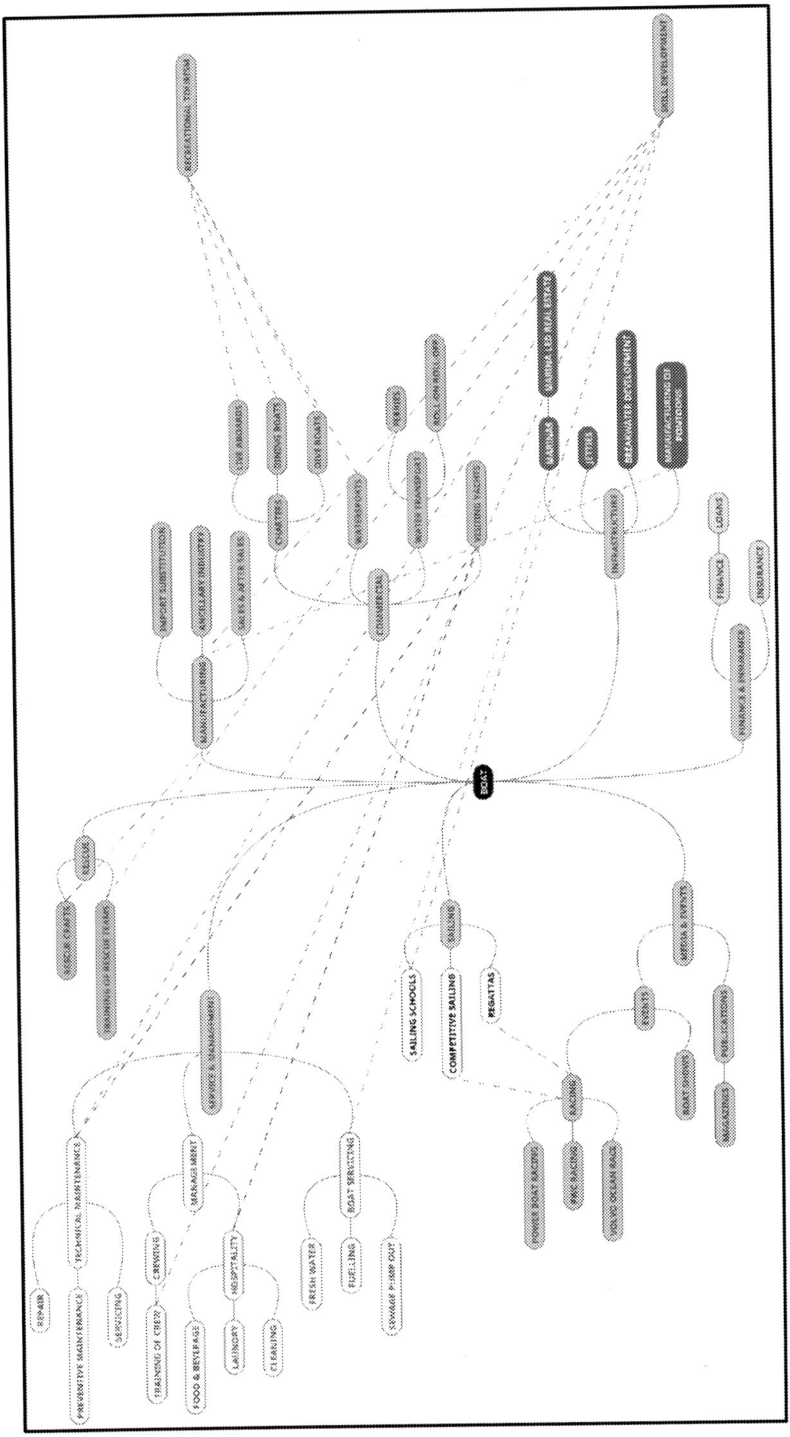

Figure 1. Creation of employment and opportunities for skill development *(Courtesy: Radhika Jhaveri)*

OPPORTUNITIES	THREATS
- Opportunity to create a comprehensive environmentally sound boating and nautical policy - Arguably the last large boating and nautical tourism market in the world - Growing number of high net-worth individuals - The success of India's first marina in Kochi has demonstrated the ability of India to attract international yachts - More sailing events (e.g. Volvo Ocean Race) helping India find a place on the international boating map - Set up Marina-led waterfront development - Sailing charters - Coastal and Island cruising - Dive Boats and Live-aboard - Employment opportunities with projects surrounding the boating-related infrastructure - A new avenue of transport - Need for quality recreational options	- No progress in terms of infrastructure for safe berthing of boats - No infrastructure for safe lifting of boats - No rescue services - Lack of skilled technical manpower for servicing of sophisticated engines and equipment - Archaic Telegraph Act making importing Radars and VHS Radios impossible which are integral for safe navigation - Law enforcement disallows charters, sighting security fears. - no commercial charters' permission - Lack of bank finance - No safety standards - No environment safeguards - Unregulated sectors that can lead to accidents and mishaps - No guidelines for boat building - No clarity on commercial cruising - High cost of ownership due to lack of marinas - Lack of skilled labour with technical knowledge - Lack of understanding of the boating and water sports sector by policymakers

Conclusion and Recommendations

The Indian boating industry got a good start with the inaugural Mumbai International Boat Show and this was evident in the 2008 and 2009 shows with a threefold increase in participants and business over INR 500 crores being transacted. However, since then a combination of factors has practically stalled the growth for at least four reasons.

First, the effect of the global economic meltdown was felt in India in mid-2010 and the slide was further accelerated by the slowdown in the Indian economy. Further, there was an increase in import duty from 26 per cent to 49 per cent. Second, the lack of boating infrastructure, jetties (to access the water), marinas (for safe harbour of watercrafts), etc. means that the cost of maintenance has increased quadruple-fold. This, coupled with the doubling of duties, became a deterrent for potential boating enthusiasts. Third, in the absence of a national boating policy, the states are applying own rules and regulations and some are even enforcing the 1918 Steam Shipping and Inland Vessel Act to yachts and personal watercraft, i.e. jet skis. It is difficult to obtain clearances to install radars and VHF radios which are essential for safe navigation because of the outdated Indian Telegraph Act. This has led to confusion and the disregard of safety and standards, all of which is evident in the quality of local, unregulated boat building. Manufacturers following international boat building practices are priced out by the high import taxes on components and local taxes in the form of excise and VAT. Fourth, the lack of boating knowledge among individuals and water sports operators, corporate entities and government, results in incorrect specifications, leading to incorrect purchase decisions, inefficient local policies, increased cost, badly authored tenders, etc. In turn, this leads to sacrificing safety and environment standards, graft, incorrect operations and maintenance and increase in overall cost of ownership and dissatisfaction.

As far as the recommendations are concerned, it is important to correct the listed anomalies. The Indian economy is again on the upswing growing at close to 8 per cent per annum, and the government needs to act swiftly to support and encourage the growth of this sunrise industry.

A comprehensive national policy is needed on boating, covering all aspects of the industry from registration of watercrafts to policy guidelines, safety standards and incentives for manufacturing, guidelines for environmental protection, particularly for the pristine islands groups of the Andaman and Nicobar and Lakshadweep atolls. It is imperative to rationalise import duties, local taxes and indirect taxes and issue clear policy guidelines which benefit both operators and users and open up new and affordable recreational opportunities for the people. The government could consider eliminating or reducing import duties for "personal watercrafts" and this can potentially have a direct impact on the water sports operators leading to more jobs and low-cost recreational opportunities. Similarly, the development of a clear policy for charter operations will open up the experience of boating to more people at an affordable cost. Finally, the government must also frame clear guidelines on infrastructure development by way of single-window clearance for jetty and marina projects and rationalise the draconian CRZ laws to boost waterfront growth.

It is important to encourage the industry by recognising it as one. The present government has, for the first time since independence, identified water transport as a key area of development and this policy must be broadened to include a national policy and guidelines on boating and nautical tourism to boost the industry.

REFERENCES AND NOTES

1. CNN http://money.cnn.com/2016/04/08/news/economy/india-population-skills-gap-education/
2. Superyacht Business, http://www.superyachtbusiness.net/marinas/china-plans-over-15000-more-marina-berths-6915
3. *Economic Times*, http://economictimes.indiatimes.com/news/economy/indicators/indias-growth-at-7-6-in-2015-16-fastest-in-five-years/articleshow/52522153.cms
4. Trading Economics, http://www.tradingeconomics.com/india/gdp-growth-annual/forecast
5. International Monetary Fund, www.imf.orghttp://www.imf.org/external/pubs/ft/weo/2016/02/weodata/weorept.aspx?pr.x=45&pr.y=1&sy=2015&ey=2020&scsm=1&ssd=1&sort=country&ds=.&br=1&c=534&s= NGDPD%2CNGDPDPC%2CPPPGDP%2CPPPPC & grp=0&a=
6. RBI, https://www.rbi.org.in/Scripts/WSSViewDetail.aspx?TYPE=Section&PARAM1=2
7. Press Information Bureau, http://pib.nic.in/newsite/PrintRelease.aspx?relid=138961
8. Press Information Bureau, http://pib.nic.in/newsite/PrintRelease.aspx?relid=138961
9. Department of Industrial Policy & Promotion, http://dipp.nic.in/English/Publications/FDI_Statistics/2016/FDI_FactSheet_JanuaryFebruaryMarch2016.pdf
10. World of Meters, http://www.worldometers.info/world-population/india-population/
11. CNN, http://money.cnn.com/2016/04/08/news/economy/india-population-skills-gap-education/
12. *Economic Times*, http://economictimes.indiatimes.com/articleshow/49336057.cms?utm_source=contentofinterest&utm_medium=text&utm_campaign=cppst
13. World Wealth Report, https://www.worldwealthreport.com/apwr/reports/population/asia_pacific/india
14. *Economic Times*, http://economictimes.indiatimes.com/articleshow/49336057.cms?utm_source= contentofinterest&utm_medium=text&utm_campaign=cppst
15. *The Financial Express*, http://www.financialexpress.com/economy/india-fourth-home-to-2-36-lakh-millionaires-report/198109/
16. Indian Tourism Statistics, http://tourism.gov.in/sites/default/files/Other/India%20Tourism%20 Book%20English.pdf
17. Nautica Italiana, http://www.nauticaitaliana.net/static/upload/496/4964b284e039cb2361b053f8ee138feb.pdf
18. Nautica Italiana, http://www.nauticaitaliana.net/static/upload/496/4964b284e039cb2361b053f8ee138feb.pdf
19. Lucintel, http://www.lucintel.com/recreational_boating_market_2020.aspx
20. Persistence Market Research, http://www.persistencemarketresearch.com/market-research/recreational-boating-market.asp
21. Freedonia, http://www.freedoniagroup.com/industry-study/recreational-boating-3203.htm
22. European Boating Industry, http://www.europeanboatingindustry.eu/facts-and-figures
23. Superyacht Business, http://www.superyachtbusiness.net/marinas/china-plans-over-15000-more-marina-berths-6915
24. Thailand Board of Investment, http://www.boi.go.th/upload/content/brochure2012_yacht_90230.pdf
25. Superyacht Business, http://www.superyachtbusiness.net/marinas/china-plans-over-15000-more-marina-berths-6915

5

Deep Seabed Mining and the Blue Economy in the Indian Ocean

G.S. Roonwal

"Mineral resources" include all mineral deposits, both known and those that remain to be discovered, that are economically exploitable now, as well as in the future. "Reserves" are a part of the known resources, the extraction of which is economically viable in the existing situation. In anticipation of the commercially successful deep sea mining, minerals like manganese nodules and seafloor sulphide may be called as "potential reserves". Strictly, these need to be called as resources until the economic viability of extraction has been established. There is no doubt that if it is financially attractive, and if other aspects like legal regimes are agreeable, the cost of extraction would be competitive and the deep sea mining of minerals could make a big contribution to the supply of metals. Deep sea mineral resources are considered as the "common heritage of mankind". Therefore, if these have to be mined, it is important that deep sea mining is developed in a way that the entire society reaps the benefit.

Introduction

Since historical times, the access to hydrocarbons, minerals and other such resources has been an area of conflict. Not all countries possess within their boundary, all the resources that are required for industrial growth and development. This results in a need for access to these resources. States look at different sources and locations and explore the possible means to access them, and several wars have been caused due to the desire for access to these resources. In the present times with free trade and movement, this has been reduced but scientific expeditions and exploration are nevertheless in progress. When the seafloor manganese nodules were first discovered on board the *HMS Challenger* in 1891, the event almost went unnoticed from the resources' point of view, though its significance was mentioned in the cruise report by Murray and Renard.[1]

When Mero[2] highlighted the economic value and showed the limitless resources of copper, nickel and cobalt in the manganese nodules lying on the seabed – the world attention was drawn to a commercial interest in mining these nodules. About this time, the United Nations (UN) was also debating on the laws of the seas for the international community. Mero's work added a new and different dimension – a sort of bonanza on the seafloor which could be extracted and refined. As a consequence of this overestimation of resources, a big hype was created for the access to these resources. This hype led to a big setback to the deliberations of the UN Law of the Sea Conference (UNCLOS). It also created a clear division between the technologically advanced nations and the developing world as to the access and benefit of metals from the mining of nodules. The UNCLOS continued as a protracted international negotiation which resulted in a set of compromises. The most difficult issue was how to mine "the common heritage of mankind", who owns what; procedures to grant permits to explore, mine sites, and processing for metal recovery; who invests and what are the economic returns?

The UNCLOS deliberations were focused on nodule mining and rules were drafted to reflect these concerns. However, in 1979, when for the first time "black smoker" chimneys were discovered in the East Pacific Rise in the North Pacific Ocean, showing that these present-day ore-forming chimneys have rich metal value (Zn, Cu, Au, Ag), as well as the fact that the water surrounding the plumes supports rich fauna – science made a big discovery. This ore formation process can be explained within the broad concept of plate tectonics. Consequently, new ores have been added to UNCLOS list of resources.

Historical Perspective

Some 410 years ago, in 1608, Grotius wrote a seminal legal work titled *Mare Liberum*, and argued that the sea is a common property of all people. Being boundless, it cannot be an object of any one state's exclusive possession, particularly for navigation or fishing. It is suited for use by all people. Over the last 300 years, since the establishment of the regime of territorial sea in the 18th Century, until the middle of the 20th Century, oceans were considered open for unrestricted use on the principles of "narrow territorial sea and wide high seas". At that time due to the needs of international trade and traffic, it was thought that maritime navigation should be free. It was also thought that fishing should be free because the fish were inexhaustible, and reproduced according to natural laws. In addition, it was a widespread myth that the oceans had an infinite capacity to cleanse themselves and it was believed that there is no risk from waste and pollutants that are dumped into the ocean. The four general conventions of 1958 were thus formulated on the basis of these ideas of freedom of the sea, particularly the freedom of navigation, fishing and overflight. However, the traditional concept of the order of the sea is in the process of change. On the one hand, it reflects the need for the establishment of a new legal regime keeping in mind new ways of using the ocean and on the other hand, this new trend is an attempt to rectify the weakness in the existing regime, under which the ocean resources would be utilised.

It needs to be pointed out here that it would be important not to overlook the fact that in addition to increasing marine pollution from land-based sources, the oil pollution at the sea is caused by accidents involving tankers and oil-drilling wells which pose a serious problem. All countries agree to the need to conserve the resources and preserve the marine environment. The idea that overfishing should not be allowed is now well acknowledged. As regards the mineral resources, a gentle warning have been sounded that land-based minerals will be depleted in the near future. Given this view, the mineral resources on the seabed are not only important but perhaps may be the only resources left for the humankind in the future.

The third UN Conference on the Law of the Sea (UNCLOS-III) has made efforts to work out a new order of the sea which would meet these emerging needs. The UNCLOS-III covers the interest of all countries and is equitable. It aims to contribute to the balanced growth of the expanding world economy. The UNCLOS-III thus fulfilled such needs not by negating the traditional freedom of the sea but by harmonising freedom and access to resources. One challenge of the deep seabed mining of minerals is by whom and what methods are the best suited for mining, within the principle of the "common heritage of mankind". Analysing the UNCLOS deliberation, two points emerge. The developing countries took the position that since the deep seabed minerals are the "common heritage of mankind", an international organisation should play a major role in the mining of these minerals. The industrialised countries, on the other hand, argued that the most rational way is for the private enterprises to undertake such development, and return a part of the profits to the international community. From this viewpoint it was advocated that the international organisation should not impose restrictions on the activities of private enterprises. The real negotiation of the UNCLOS started with such basic differences. As the painstaking negotiations and the process of drafting the text - first text, then a revised text – and finally a single negotiating text concluded, eventually an informal composite negotiation text emerged. During this process, large changes were incorporated at each stage. After the single negotiating text emerged, the process of licensing proposed by the industrialised countries was not accepted, and a formula of direct mining by an international organization – the ISA's "Enterprise", was instituted. Later, at the 4th session of UNCLOS, after a counter-move by the industrialised countries, changes were effected in the so-called revised text. A single negotiating text suggested that in addition to direct mining through 'Enterprise'-ISA, private consortia may participate in mining. Thus a parallel system emerged, though the developing countries opposed it. The 5th session of the UNCLOS ended with a repetition of the basic views of both sides. However, in a new move, the US Secretary of State Dr H. Kissinger put forward a proposal concerning financial help for the 'Enterprise'-ISA, the transfer of technology and production control. This did not resolve the deadlock and in the 6th session, the "Evensen Text" was announced which accommodated the position of the developing countries while making some concessions to the industrialised countries. In the 7th session, the UNCLOS was concluded and remains in force since November 16, 1994. In essence, it was agreed that the ISA would exercise control over the deep-sea mining and the 'Enterprise' ISA shall engage in mining activity. However, the question which has to be examined is that does the 'Enterprise' have the necessary personnel, capital and technology, in order to make deep-sea mining truly workable?

At the end of the Cold War, a new dimension has emerged. The nations aspiring to be world powers have focused on the use of the sea not as a "common heritage of mankind", but as an area of influence. In the name of resource exploration/mining, a sort of competition has emerged in the Indian and the Pacific oceans, but it is more evident in the Indian Ocean.

Deep Seabed Mining Resources

The Deep Seabed Mining Resources are spread over a large area. Table 1 shows the typical contents (by wt percentage) of selected metals from seafloor poly-metallic sulphide samples and related ancient massive sulphide ores.

Table 1. Typical contents of selected metals from seafloor poly-metallic sulfide samples (by wt in percentage)[3]

Location	21°N Active Vents	Cyamex Area	Modern Juan de Fuca Ridge Northern	Modern Juan de Fuca Ridge Southern	Modern Juan de Fuca Ridge Axial Seamount	Mid-Indian Ocean Ridge CpSpl	Ancient Cyprus	Ancient Canadian Precam-brian
Zinc	32.3	40.8	6.3	54.0	19.2	0.14 31.10	0.2– 0.5	5.0
Copper	0.8	0.6	0.5	0.2	0.13	31.6 0.71	2.5– 4.0	2.0
Lead	0.3	0.05	0.1	0.3	0.4	-		0.0
Silver (ppm)	156	380	30	260	288	44	39	50
Gold (ppm)	0.17	-	0.08	0.13	-	890	0.3	181.76

Table 2. Nodule Composition (Average in Per Cent)[4]

	Pacific Ocean	Atlantic Ocean	Indian Ocean
Mn	19.78	15.78	15.10
Fe	11.96	70.78	14.74
Co	0.335	0.318	0.230
Ni	0.634	0.378	0.464
Cu	0.392	0.116	0.294

Seafloor sulfphde offers a better option; the hydrothermal mineral deposits or poly-metallic sulphides and their associated metalliferous sediments consist of sulphides of iron, copper and zinc, oxides and silicates of iron, and oxides of manganese. These are formed by hydrothermal processes associated with seafloor volcanic activity. Most of the occurrence has been reported on the mid-ocean ridge system (including the Red Sea) and to a lesser extent, on convergent plate margins[5,6,7]. Most of the deposits on mid-ocean ridges occur in the International Seabed Area. In contrast, most of the convergent margin deposits fall within the Exclusive Economic Zone (EEZ) of the states.

India's Position on Deep Seabed Mining in the Indian Ocean

India has faced the dilemma of access to resources, which has changed from resource diplomacy to resource competition. Table 3 shows the licenses which have been issued for seafloor mineral exploration in the Indian Ocean.

In September 2016, India signed the contract for exploration of seafloor sulphides in the Mid-Indian Ocean Ridge, with the ISA. This allows India to go for a detailed exploration of the area for the next 15 years.

The other important parameter is concern for the environment. This has been seen in land mining and it is evident that mining damages the environment. The challenge is therefore how to overcome, contain and bring mining to an acceptable level which the society readily accepts. After all, the metals produced are for the benefit of the people and help improve their quality of life. Without wealth creation, development cannot happen; therefore, the resources of all types, whether on land or seabed, have to be used for the benefit of the society.

Economics of Deep Sea Mining

It is possible to develop resources lying on the seabed at several thousand metres' depth due to advances in scientific methods, engineering and technology. It is argued that this is a manifestation of human creativity and determination, the intellect and desire to improve the quality of life of people.

In order to mine the seabed minerals, it would be necessary to build on the principles proposed in the UNCLOS documents. It is expected that there would be close cooperation amongst the nations since deep sea resources are the "common heritage of mankind". However, there are some nations which have the technical know-how and capital for developing deep sea minerals, while other nations do not have these resources at their disposal. Hence, it is expected that there will be a conflict of interest over these issues. Some hard issues which may require discussion and resolution are as follows:

(i) The system of exploration and mining
(ii) Production policies
(iii) Transfer of technology
(iv) Financial arrangements
(v) The corporation, power and function of the ISA/Enterprise

At this juncture, it is useful to have a quick view on the economic dimension of nodule mining. The generally accepted minimum criteria

Table 3. Licenses for Seafloor Mineral Exploration in the Indian Ocean

Contractor	Draft entry into the contact	Expiry	Sponsoring state	General location of the exploration area under contract	Type	Depth	Area (km^2)
Govt. of India	25th March 2002	24th March 2017	India	Indian Ocean	Poly-metallic nodules	5000-5700	75000
China Ocean Mineral Resources and Development Association (COMRA)	18th Nov. 2011	17th Nov. 2026	China	South West Indian Ridge	Poly-metallic sulphide	-	10000
Govt. of Republic of Korea	1st May 2014	30th April 2029	Korea	Central Indian Ocean Ridge	Poly-metallic sulphide	-	10000
Govt. of India	26th Sept. 2016	25th Sept. 2031	India	Central Indian Ocean	Poly-metallic	3000	10000
Federal institute of Geosciences and Natural Resources of Germany	16th May 2015	15th May 2030	Germany	Central Indian Ocean	Poly-metallic sulphide	2600-3300	10000

regarding nodule metal grade, metallurgical processing efficiency, annual nodule production, annual metal production of each mining block can be determined using assumptions in Table 4.

Table 4. Assumptions for Production of Metals from Deep Seabed Mining

Metal grades	Metallurgical processing efficiency	Expected production from each block
Ni – 1.3%	Ni, Cu, and Mn 90%	Ni – 35100 metric tons
Cu – 1.1	Co 70%	Cu – 29700 metric tons
Co – 0.25		Co – 4200 metric tons
Mn – 27.0		Mn – 729000 metric tons
		Total – 3 million metric ton per annum

Estimate of the growth in demand for metals: Ni – 2.6%; Cu 3-5%; Co 3-6%.

The projections of total annual production in the future and the likely range of impact on the world metal markets can help compute the economic value of deep seabed metals. As noted earlier, the preliminary estimates of Mero (1965) were inflated; since then, subsequent serious deliberations have lowered the production estimates from trillion to million tonnes.[8] It can be estimated that about 15 per cent of the total area of the world's ocean floor (362 million km^2) is covered by nodules. Surprisingly, this estimate is consistent with the evidence provided by the *HMS Challenger* expedition. It is important to keep in mind that many nodules contain only very small quantities of Ni and Cu. Similarly Mn nodules contain only very small quantities of Ni and Cu. Therefore, to making nodule mining competitive, costs have to be greatly pruned.

It is also suggested that the costs and the resources involved can be estimated more accurately by dividing the work into the following:

(i) Defining the mining area
(ii) Quantity of minerals ore and quantities of metal extracted
(iii) Quantity and stall of grade
(iv) Scale and quantity of production
(v) Mining technology and evaluation method to be used
(vi) Type of mining plants
(vii) Method of transportation
(viii) Extraction of technology
(ix) Environmental problem related to these
(x) Feasibility of mining

Conclusion

The competition for seabed exploration and access to resource is a welcome development. It is hoped that this competitiveness shall encourage mutual cooperation towards the production of metals which will lead to growth and development. It may be noted that in the ISA, matters would be decided by a two-third majority, and since developing countries form two-third of the world's countries, the industrialised countries would be in a position of minority. In such a situation, there are bound to be issues which can be raised to protect the position of the industrialised countries. The last few years has seen a radical change in the position of the Indian government. The general trend has been towards the liberalisation of the economy and to lower the traditional interventionist policy. The major priority of the administration has been to increase the contribution by the private industry in science and this opens up an opportunity for mutual cooperation with other nations.

REFERENCES AND NOTES

1. Murray J. and A.F. Renard, 'Deep-Sea Deposits: Report on the Scientific Results of HMS Challenges 1873-76', HMSO, 1891.
2. Mero, J.L., *Mineral Resources of the Sea*, Amsterdam: Elsevier, 1965
3. Data source: Adapted from various sources.
4. Cronan, D. S., 'Underwater Minerals', *Academic XV + 362*, 1980.
5. *Economic Geology*, 'A Special Issue on Seafloor Hydrothermal Mineralization: New Perspective', Vol .88(8), 1993.
6. Rona, P.A., K. Boston, L., Laubier and K.L. Smith (eds.), *Hydrothermal Processes at Seafloor Spreading Centres*, NY: Plenum, 1983.
7. Scott, S. D., 'Seafloor Poly-Metallic Sulfides, Scientific Curiosity or Mines of Future?' in *Marine Mineral* P. Teleki + M.R (ed).
8. Glasby, G.P. (ed.), Marine Manganese Deposits, Elsevier, 1977.

6

Marine Spatial Planning and Ocean Ecology: Prospects and Lessons for the Indian Ocean

M. Imali Udeshika Manikarachchi

Marine Spatial Planning (MSP) is a strategic tool to plan and manage conflicting ocean uses and their interactions with the marine ecosystems. The MSP allocates space for marine-based industries and activities to fulfil the demand for marine-based goods and services while facilitating environmental conservation by imposing regulatory and management measures. It reduces cross-sector conflicts, enables proactive decision-making and safeguards valuable ecosystem services. The MSP approaches are long term with continuous learning and adaptations, making the process more dynamic. This chapter discusses the role of MSP in Ecosystem-Based Management (EBM), the characteristics and processes of an effective MSP approach and the zoning mechanism of MSP, taking Australia's Great Barrier Reef Marine Park as an example. It further looks into the details of the significance of the Indian Ocean and prevailing potentials and challenges to implement the MSP initiatives in the region.

Introduction

Marine resources vary from animal proteins to deep-sea minerals and humans rely on them for their subsistence. Oceanic routes facilitate the most efficient means of transportation in the world and the scenic beauty of beaches attracts tourists and thousands of locals thus generating direct and indirect employment and a source of income. Oceans are home to the largest ecosystems on the planet, which provide significant ecological services. Coral reefs, sea grass and algal beds, deep-water and coastal habitats sustain marine and coastal biodiversity. In fact, they are the feeding and breeding grounds for fish, crustaceans and many other vertebrates and invertebrates. These coastal habitats line the coastal belt and protect it from shoreline erosion and natural disasters such as tsunamis. Oceans are the main driving forces of global climate and a major sink of atmospheric CO_2. The deep oceanic currents drive the conveyer belt of thermohaline circulation, which makes life possible in and beyond the tropical regions.[1]

Modern technology has made the oceans more accessible and the concept of the 'Blue Economy', which promotes a sustainable utilisation of marine resources to achieve economic, social and environmental benefits, has gained worldwide recognition.[2] However, its practical applications might not exactly follow the expectations, since the quality of the marine environment is rapidly declining due to the increasing anthropogenic pressures all over the world,[3,4] Oceans have become increasingly crowded with numerous floating and submerged industries that extract resources at a rate higher than their rate of generation.

Marine Ecosystems and Services

Oceans account for 71 per cent of the earth's surface, making it a 'Blue Planet'. They are comprised of the largest and most diverse ecosystems in the world, which facilitate the interactions among living marine organisms and their surrounding abiotic environment. They range from coastal habitats to open ocean and deep-sea habitats with different extremes of ecology and biogeochemistry. Marine ecosystems such as coral reefs, sea grass and algal beds, mudflats, mangrove forests, estuaries, lagoons and salt marshes are common in the coastal zone while the water column of the open ocean itself supports a vast array of life forms. Deep-sea ecosystems, such as hydrothermal vents and cold seeps, also form unique livable environments for chemoautotrophs despite the frigid temperatures, high pressure, fading sunlight and anoxic conditions prevailing there.[5,6]

These ecosystems provide benefits in the form of ecosystem services, defined as the flows of benefits from nature to people.[7] Marine ecosystem services generally include climate regulation, carbon sequestration, nutrient recycling, sheltering fishery resources, protection from coastal erosion and assimilation of wastes and other support mechanisms. Other benefits provided by the oceanic environment may include provisioning services such as goods (fish, seaweed, minerals, energy) and cultural services such as natural beauty and recreation. These services are provided as a result of complex aggregations of different functions of ecosystems, which once degraded, reduce the quality of those services.[8]

Significance of the Indian Ocean Region

The Indian Ocean (IO) accommodates a rich diversity of various pelagic and benthic marine ecosystems. It includes 246 estuaries and a large number of minor estuaries, besides coastal lagoons and backwaters.[9] Tropical climate conditions of the IO enable the survival of 30 per cent of the world's coral cover.[10] The Hooghly estuarine system in India and the Brahmaputra estuarine region in Bangladesh, sustain the Sunderbans, which are among the largest mangrove forests in the world.[11] The combined length of the coastlines of all coastal states of the IO littorals exceeds 66,000 km and the average area of the intertidal zone covers around 3,000 km^2. The Indian Ocean generates an annual yield of 8 million tonnes of capture-based fisheries and 23 million tonnes of culture-based fisheries, providing many people with affordable animal proteins and livelihood. Being landlocked to the northern side, the IO experiences a unique pattern of monsoons, driven by a reversal of the direction of the wind and surface currents. Also, the mean annual precipitation varies greatly between the western and the eastern side of the IO, forming prominent salinity variations. These two forces have a significant influence on the IO's productivity and biodiversity. Moreover, nine Large Marine Ecosystems (LMEs) are functioning in the IO.[12]

Threats to Marine Ecosystems of the Indian Ocean

The myth of oceans being unlimited reservoirs of resources has changed and records of overexploitation of oceans are found all over the world. Since it is only one global ocean without physical boundaries, the impacts may more or less result in detrimental effects to the entire system. A large portion of the marine ecosystems of the IO is now at the point of degradation due to this rising pressure, which has made the coastal regions and their communities more vulnerable to environmental risks.[13] The threats to the marine ecosystems of the IO are briefly enumerated in the following paragraphs.

Overexploitation of Marine Resources

Overexploitation threatens food security and biodiversity and marine fisheries resources are mostly affected by overexploitation. If fish stocks completely collapse, it is unlikely that they would ever be restored to the previous levels. In the Western IO, most of the fish stocks are fully exploited where 24 per cent of the fish stocks have no room for further expansion and in the Eastern IO, the major fish stocks are being overfished due to an uncontrolled open access.[14] In the Eastern IO, resource degradation is very much related to population and the high unemployment rate of the coastal states.[15] Another reason for overexploitation in this area is poaching, due to lack of monitoring, control and surveillance capacity.[16] According to the Trans-boundary Diagnostic Analysis of the BOBLME project, overexploitation of the marine living resources has been identified as one of the three major priorities to be addressed, others being habitat degradation and pollution.

Pollution and Degradation of Marine Ecosystems

Coastal pollution is heavily influenced by excessive amounts of nitrogenous chemicals, which are introduced to the rivers by agriculture. In addition, untreated sewage may find its way to oceans through river water discharges. These pollutants result in hypoxic dead zones, which are harmful for marine living resources.[17] In India, 18,240 million litres of domestic sewage are discharged into the rivers every day.[18] Major stressors of the Western IO include greater volume of untreated sewage and nutrient runoff released into the sea.[19] The busy shipping lanes in the Indian Ocean threaten the region through invasive species, discharged along with the ballast water.[20] Especially in the high seas, deep-sea fisheries resources of the Southern IO are being exploited by bottom trawling and significant impacts have been recorded in the form of fish and coral bycatch.[21]

Coastal tourism plays a significant role in the economy of many coastal countries' as their scenic value has made them popular tourism hotspots. The development of tourism activities, coupled with coastal urbanisation, gradually creates a huge impact on these sensitive ecosystems.[22] These are also disturbed by building infrastructure that may require a clearing of the coastal vegetation. Tourists tend to bring materials such as polythene and plastics and leave them behind. They get accumulated along the coasts and are taken offshore by splashing waves and may stay there for a very long time causing numerous ecological problems.[23] The careless disposal of waste from tourist resorts causes pollution problems,[24] while irresponsible boating, swimming, diving and trampling may physically damage the fragile marine ecosystems such as the coral reefs.[25,26,27].

Climate Change

The vulnerability of the region for climate change is further enhanced by the high population growth and poverty. Among the more prominent impacts of climate change in the South Asian region are increased air temperatures, changes in the annual mean rainfall and its frequencies which have been increasing over time.[28] Moreover, in the Bay of Bengal and the Arabian Sea regions, the frequency of the cyclonic events, since the 1970s, has decreased, but the intensity has increased, causing severe damage. Increased sea surface temperature is also evident in the South Asian region, which results in thermal expansion, followed by sea-level rise. There is an observed sea-level rise of 1–3 mm per year in coastal South Asia, which is slightly higher than the global average and the rate of sea-level rise also has increased, compared to the 20th Century.[29] Consequently, lowlands along the coasts are at risk of frequent flooding and Maldives is the most dramatic example.[30] Another impact of the anomalies of the sea surface temperature is the El-Niño, which is one of the reasons for declined fish stocks in the South Asian coastal waters. This is the major reason for large-scale coral bleaching in the region.

Issues of Maritime Security and Disputes

The Independent World Commission on the Oceans (IWCO) has emphasised the importance of maritime security for peaceful uses of the oceans, where benefits have to be shared by all, under an equitable public order.[31] The maritime security issues are of an international concern and may surface as military threats such as piracy and terrorism, environmental threats, non-traditional maritime security threats, IUU fishing or ecosystem destruction. Shipping in the Indian Ocean region is becoming more concentrated due to the shift in the focus of the world economy from the North Atlantic to Asia, and the interest in the resources in the IO seabed is increasing.[32] There has also been an upsurge in piracy and terrorist attacks. The region is at environmental risks associated with possible oil and chemical spills, due to the increased maritime traffic, of which the impacts could be significant and long term.

Need for Management

Garrett Hardin, an American ecologist and philosopher, suggested that jointly held natural resources are subjected to the *tragedy of the commons*, which he stated as "Freedom in common brings ruin to all".[33] This simply means that when natural resources are shared among multiple consumers, they will be eventually depleted by overexploitation. Increasing population along the coasts causes more pollution and exhaustion of natural resources at a rapid rate, threatening the ecosystems that provide ecological and economic benefits.[34] This will gradually reduce the ability of the ecosystems to provide its tangible and intangible services in the long run.

Before the 20th Century, the oceans had been explored for fishing and utilised for navigational purposes only, which raised few conflicts among the interested parties.[35] Therefore, the issues could be overlooked and addressed sector-wise. However, this approach has created difficulties in ocean management. Today, multiple maritime industries are emerging all over the oceans and they have started to crosscut and intertwine with each other. This has resulted in a competition for ocean space causing user-vs,-user conflicts and user-vs.-environment conflicts[36] and sector-wise management has become outdated and ineffective. Therefore, ocean management has to be approached in a holistic way such that all these dimensions are given proper attention.

Innovative ocean management measures, that are capable of balancing the three pillars of sustainability – economy, society and environment – have surfaced especially after the 1992 Rio Summit. This brought a paradigm shift and introduced the concept of 'Sustainable Development'. Agenda 21, a 40-chaptered document was produced as an outcome of the Rio Summit, with a dedicated chapter for the protection and the development of living resources of oceans and coastal areas. Chapter 17 of Agenda 21 notes the importance of oceans in the global life support system and the opportunities for its sustainable development. More importantly, it facilitates new integrated approaches for management measures of the oceans.[37] Moreover, the recent UN 2030 Agenda for Sustainable Development dedicates Goal 14 for the conservation and sustainable use of oceans and marine resources, where it demands strengthening the partnership among different stakeholders for a better management of the oceans.[38]

Ecosystem-Based Management (EBM) Approach

The EBM is a relatively new approach of natural resources management that is different from the traditional, less effective single sector management practices, which have often resulted in mismatched governance regimes and coordination issues. The EBM identifies the human interactions within the ecosystems and their interdependency. If an ecosystem is stressed by multiple human engagements, their cumulative influence can conflict with the proper functioning of that ecosystem. In the EBM, preserving and restoring habitat quality and maintaining the ecosystem services have to be done by a proper management of human activities in the ecosystems.[39] It requires a thorough understanding of the structure, functions and processes of the ecosystem to determine the level of human behavioral changes needed to restore the ecosystem services. Usually, the EBM is an integrated, collaborative and adaptive approach and is to be supplemented by scientific knowledge and effective monitoring.[40]

Role of Marine Spatial Planning in EBM

Marine Spatial Planning (MSP) is a public process that is capable of analysing and allocating the spatial and temporal distribution of human activities in marine areas.[41] It is also a tool for improved decision-making and delivers an effective ecosystem-based approach in managing human activities in the marine domain.[42] The MSP enables the recognition of all activities carried out in an area and their linkages with the marine ecosystems, thereby allowing a long-term sustainability of both the human engagements and the natural processes.

Many European countries consider the MSP as an operational instrument to implement EBM.[43] In most cases, ecosystems are localised and place-based, which can be influenced by the existing cultural, economic, and political attributes.[44] For a better implementation of the EBM, it is crucial to understand the ways of integrating these components with the existing ecological knowledge and perceptions. This provides policymakers and managers with an accurate picture of the ecological, socio-economical and judicial dimensions that interact in a delineated marine area. Long et al. (2015) have provided a list of 15 key principles that define the concept of EBM. According to them, the consideration of ecosystem connections, adaptive management, appropriate spatial and temporal scales, use of scientific knowledge, stakeholder involvement and integrated management are the most important principles to be expressed in an EBM plan. According to Ehler and Douvere (2007), characteristics of MSP shall include integrated cross sectors, strategic long-term goals, active stakeholder participation, adaptive management, balanced economical, ecological and social pillars and spatial demarcation.

Benefits and Characteristics of MSP

The MSP is capable of delivering significant ecological, economic and social advantages, which are shown in Table 1.

Table 1. Advantages of Marine Spatial Planning[45]

Ecological Advantages	Identification of ecologically sensitive areas. Identification of areas incompatible with human engagement. Conservation purposes (e.g.: Marine Protected Areas)
Economic Advantages	Identification of compatible and conflicted sectors. Identification and encouragement of future potentials. Enhanced maritime safety Efficient space and resource allocation
Social Advantages	Identification of culturally important maritime space. Better access to open ocean Improved information exchange and communication

(a) *Integrated cross sectors:* Sectors and different levels of governing systems can be integrated, generating an overview of all the economic, environmental and social objectives to be achieved by the planning process.

(b) *Strategic long-term goals:* A long timeframe is allocated for plan designing, implementing and monitoring, with possible revisions.

(c) *Active stakeholder participation:* Different stakeholders and their involvement with the marine environment have to be properly understood at an early stage and sustained throughout the process. Stakeholder participation enhances the communication, support, trustworthiness and sense of ownership among the parties who are about to be impacted by the respective MSP. It is important to achieve the right mix of expectations to minimise conflicts through stakeholder consultation. Stakeholders may engage through public meetings, written submissions, public polling, advisory groups and sector-based group discussions.

(d) *Adaptive management:* Effectiveness of the management measures shall be constantly improved, updated and evaluated. Revisions, reconsideration and redesigns based on scientific recommendations are essential for adapting to changing conditions.

(e) *Balanced Economical, Ecological and Social pillars:* Recent literature reveals that the EBM can be considered as the best way of delivering sustainable marine ecosystems goods and services.[46] In many parts of the world, the sustainability concept has shaped the direction of marine policies. According to the 'European Commission's Roadmap for MSP', the objective of the MSP is to "balance sectorial interests and achieve sustainable use of marine resources". The MSP provides a platform to identify areas for conservation and integrates the uses of maritime space, which are the 'hard' and 'soft' concepts of sustainability, respectively.

(f) *Spatial demarcation:* One of the important steps of the MSP is the delimitation of boundaries by generating zoning maps, which reflect place-based geomorphological, ecological and ocean use information. This space allocation helps managers to assess the cumulative human impact on the environment, existing and potential conflicts, governance structures and room for future development, thereby facilitating strategic planning.

Processes of MSP

The whole process of the MSP must be long term, with an output of a comprehensive spatial management plan for different uses of the sea. Literature suggests that the MSP consists of at least three basic processes,[47] which encompass ten sub-processes.[48] Plan making, plan implementation, and monitoring and evaluation are the three basic processes that are adaptive and repetitive over time. Under these three basic processes, there are ten quasi-linear sub processes.

(a) *Identifying need and establishing authority:* The understanding of existing or anticipated needs helps in setting goals and provides guidelines to stay focused during the entire process. There should be authorities for both planning and implementing purposes for the MSP. The most common way is the establishment of a separate authority for planning while the existing institutions carry out implementation. The existing legislation can be revised, new legislation can be created or new provisions can be added to the upcoming legislation in order to empower the planning bodies of the MSP. A central authority usually takes the responsibility of plan implementation or existing single-sector management authorities remain responsible for implementation, consistent with the major plan. In some cases, a mixed approach of these two options is followed.

(b) *Obtaining financial support:* Since it is a government initiative, the investments are from direct budget allocations. In case of insufficient funding, support can be requested from regional and international funding organisations, the private sector, donor organisations and revenues from sea use.

(c) *Organising the process through pre-planning:* Pre-planning puts everything in order prior to developing the actual plan. In this phase, the process involves forming a multidisciplinary planning team, making a working schedule, setting up boundaries and timeframe, specifying principles, goals and objectives, assessing risk and developing contingency plans.

(d) *Organising stakeholder participation:* Expectations, opportunities and conflicts of ocean uses can be mutually discussed by bringing the stakeholders onboard during the process of the MSP. Stakeholders may range from scientists to fishermen with traditional knowledge and any other persons who may be affected by the proposed plan. However, the right number of relevant stakeholders should be considered at the right time, which may otherwise cause many distractions.

(e) *Defining and analysing Existing Conditions:* The collation of reliable and updated information is possible from scientific literature, scientific opinion or advice, government sources, local knowledge and direct field measurements. Database, inventories and maps of ecological features (spawning habitats, feeding ground, sensitive ecosystems, etc.) oceanographic features (bathymetry, currents, etc.) and human activities (commercial fishing, offshore aquaculture, shipping lanes, etc.) have to be created over time, in order to assess the conflicts and compatibility of sea uses and the impact on environment. These data can be obtained using remote sensing, which then can be managed and presented in the forms of maps using GIS.

(f) *Defining and analysing future conditions:* In this step, future trends and alternatives of spatial sea use under the influence of new management are foreseen. Based on different goals, several spatial sea use scenarios can be produced, out of which the most suitable one can be selected. Historic trends and drawing programmes can project future trends.

(g) *Preparing and approving the spatial management plan:* A spatial management plan describes the boundaries of the area, the timeframe, management goals, measures and financial plan. The identification of alternative sector-based spatial management measures (e.g.: VTS schemes for shipping routes, safety zones for ports, closure areas for fishing, etc.) and evaluation and approval of the plan are the major events in this step.

(h) *Implementing and enforcing of the spatial management plan:* Plan implementation requires an effective execution of the plan with possible improvements and adaptations. An integrated vision of achieving multi-sectoral goals, together with environmental conservation, is produced in the form of a policy statement. Compliance from the stakeholders' side shall be assured through education and providing technical assistance. Taking legal actions against noncompliance and violations enforces the provisions of the plan.

(i) *Monitoring and evaluation performance:* Performance indicators (the measurements of how well the plan has achieved its objectives) have to be identified. The required baseline data needs to be collected by direct observations, interviews and surveys.

(j) *Adapting the spatial management process:* Based on the results of monitoring and evaluation, the management plan may be modified. New research possibilities can be identified based on the future information demand.

MSP in Practice: Case of the Great Barrier Reef Marine Park of Australia

Australia's Great Barrier Reef (GBR) is the world's largest coral reef ecosystem with a spectacular economical, ecological, social and aesthetic importance. It has been declared as a UNESCO World Heritage site. Not only the corals, but the reef itself houses numerous other ecosystems such as sea-grass beds, mangroves and benthic habitats for algae and sponges and deep ocean troughs.[49,50] During 1960 to 1970, continuous human impacts such as oil drilling, limestone mining, pollution from shipping and land-based sources, fishing and tourism were the major driving forces of reef degradation. The importance of conservation was highlighted with the establishment of a marine park in 1975, encircling and protecting a large part of the GBR. The Great Barrier Reef Marine Park (GBRMP) has been established under the GBRMP Act, 1975 with an overall goal "to provide for the long-term protection, ecologically sustainable use, understanding and enjoyment of the Great Barrier

Reef through the care and development of the Great Barrier Reef Marine Park".[51]

Though zoning had only been used as a terrestrial spatial management technique when the GBRMP was first declared, the latest amendments of the Act have accepted the application of zoning as the core element of its MSP and other measures such as permits, mutual agreements and education.[52] Zoning has been used for the management of the GBR since 1981.[53] Zoning is a spatial planning tool that is similar to a town planning scheme, which separates conflicting uses from sensitive, ecologically valuable or recovering areas.[54] According to the GBRMP Act, the objectives of zoning of GBRMP include conservation, regulation of the uses, regulation of exploitation activities, reservation of areas for public appreciation and preservation of the natural state. The presence of particularly vulnerable reef areas, species or communities and the prevailing issues are being addressed in plans that complement the major zoning plan. Table 2 shows the states of activities in the zonal plan of the GBRMP.

Table 2. GBRMP Zoning Plan[55]

Zone	Permitted	Limited	Prohibited	Remarks
General Use Zone	All reasonable activities including shipping, trawling,		Mining, oil drilling, commercial spear fishing, spearfishing with SCUBA	
Habitat Protection Zone	All reasonable uses including most commercial and recreational activities		Shipping and trawling and activities not allowed in the "General Use Zone"	"No Trawling Zone"
Conservation Park Zone	Appreciation and recreational use	Line fishing	Spear fishing and collecting, activities not allowed in "Habitat Protection Zone"	"Limited Fishing Zone"
Buffer Zone	Pelagic trolling, appreciation and enjoyment of areas in their relatively undisturbed state.		Activities not allowed in "Conservation Park Zone"	"Trolling only Zone"
National Park Zone	Appreciation and enjoyment of areas in their relatively undisturbed state		Fishing and all forms and extraction	"No take zone" (Green Zone)
Scientific Research Zone	Exclusively for marine research			
Preservation Zone	Exclusive scientific research		Entrance except in an emergency	"No go zone"
Commonwealth island zone	Use of areas of the Marine Park on Commonwealth Islands			

In 2004, the initial zonal plan was revised to increase the areas of "Habitat Protection Zone" and "No Take Zone" to enhance the biodiversity of the GBRMP. Significant attention on stakeholder participation was paid during the "initial zoning" by forming advisory committees comprised of local resource users and sector-based experts. Maps defining locations and written objectives are mandated in each zone. The zoning provisions state whether the activities that occur in the zones require permits/notifications or not. The activities in each zone are managed by sector-based agencies in order to align with the major zoning plan. In addition to zoning, some other non-spatial management measures are applicable such as best management practices, impact assessments, public education and management charges. The Act has been amended to require the GBRMP Authority to produce an "Outlook Report" every five years to reveal the overall condition of the GBRMP and to understand the future changes of the plan. This reflects the adaptive nature of the whole process.

MSP in the Indian Ocean

The Indian Ocean region has a high coastal population density, representing 30 per cent of the world's population. Most of the IO rim countries are developing counties characterised by chronic poverty, low per capita income, struggling economies and fragmented political systems.[56] However, over the past few years, the IO region has managed to become one of the most economically and politically concerted places in the world, especially due to the presence of maritime choke points such as the Suez Canal, the Bab al-Mandeb, the Strait of Hormuz and the Strait of Malacca.[57] Being the third largest ocean in the world, the IO is home to a vast array of natural resources including, fisheries, minerals and hydrocarbons. These marine resources, if sustainably and innovatively exploited, can bring long-term prosperity to nations, the concept which is now evolving as the Blue Economy. The Indian Ocean Rim Association (IORA) has affirmed the Blue Economy as one of the top priorities in the region to generate employment and to ensure sustainable businesses and economic models.[58] It ensures regional cooperation to develop the vital areas of Blue Economy such as fisheries and aquaculture, renewable ocean energy, seaports and shipping and mineral exploration, which may further enhance the demand for maritime spaces.

These diverse opportunities require proper coordination among various stakeholders in order to share ideas on mutual ocean usage, to minimise conflicts and on regulatory issues and to develop a resilient knowledge base.

These interactions need to be embedded at policy levels to make effective decisions for the betterment of the economic, social and environmental attributes. Moreover, the economic potential of marine ecosystems collapses with increasing anthropogenic impacts that may create "dead zones", with no ecological value. Therefore, conservation, management and governance of marine ecosystems should go hand in hand with industrial applications in the oceans until it strikes the right balance.

Countries such as the US, Australia, New Zealand and many European nations have successfully incorporated regional MSP initiatives in their respective marine policies to resolve the exiting ocean governance issues.[59,60] The existing MSP programmes cover around 10 per cent of the world's EEZs.[61] The MSP is relatively a new concept for the IO region although a few countries have started to implement the process or have taken it into consideration. One example is the case of Seychelles in the Western IO, initiated in 2014. The declared goal is to sustain the long-term utilisation and health of Seychelles's EEZ by contemplating the Blue Economic opportunities, marine protection, climate change adaptation and other national strategies in a MSP programme.[62] The major sectors include commercial fishing, tourism, biodiversity conservation, renewable energy, maritime safety and non-renewable resources and zonal plans have been designed concerning the available spatial data, economic information and local knowledge of the stakeholders. The Ministry of Environment, Energy and Climate Change has the responsibility of planning with the funds provided by the Government, private and international partnerships. Seychelles has declared 30 per cent of its waters as protected areas while achieving an improved management of fisheries and tourism sectors, a transparency of decision-making and an improved understanding of the maritime spaces. Apart from that, prospects to implement MSP projects, especially in the Western IO, have been assessed by several countries such as Madagascar[63] and South Africa.[64] In the Western IO, MSP initiatives have been used to conserve the ecological integrity of the Agulhas Somali Current Large Marine Ecosystem, by declaring the Trans-Boundary Marine Protected Areas.[65] In the Eastern IO, localised regions of the Philippines and Indonesia have already taken steps to practice MSP.[66] Compared to the Western IO, literature doesn't support the existence of MSP in the Eastern IO. However, almost all the countries in the IO region have prepared Integrated Coastal Zone Management (ICZM) plans. An ICZM plan is slightly different from the MSP even though the two concepts share common grounds such as integrated approach, stakeholder participation and avoiding conflicts. The two concepts differ as the boundaries of the MSP go beyond the coastal zone to offshore areas, while the ICZM usually has been assigned with coastal attributes.

If current trends continue, the IO is going to be highly exploited by a variety of resource users in the near future and decisions will have to be made on how to make their involvement with the marine environment sustainable.[67] Therefore, planning has to start from an early phase anticipating the future scenarios, to which the MSP is capable of throwing some light. Clearly, it needs a robust plan to benefit from future opportunities and to manage conflicting situations, which extend much further from the coastal region. The MSP can be a supplementary tool to the CZMPs of the IO-rim countries to connect the coastal regions with the offshore environment up to the high seas, providing a comprehensive platform to oversee the entire maritime domain. Institutional changes have to be made in order to authorise MSP initiatives with a necessary sector-wise coordination. The lack of intellectual attention on the MSP, and limited socio-economic and environmental data on oceans, have so far made it difficult to formulate policies that support the MSP initiatives. The situation is further challenged by inadequate capacity and by technical, legal, and institutional concerns in the implementation process.[68]

Conclusion

Natural and human-induced processes are heterogeneous in nature and coexist in the marine environment. Natural elements such as the ecosystem itself and its functions cannot be regulated or planned. Therefore, it is only the human element that can be planned or regulated in order to restore and maintain good order of the marine environment. Therefore, it is essential to take a systems approach to understand the interactions among multiple natural and human drivers and cumulative impacts generated from these integrations on the environment. An adaptive governance regime is needed in these circumstances, that enables monitoring and surveillance and can also cope with uncertain conditions that may arise in the future. The MSP provides a platform for different levels of governing bodies and stakeholders to reach multiple objectives, which makes the decision-making process more transparent, smooth and progressive.

Considering the steps to follow, the MSP can be expensive for some developing countries. Poor financial allocations, governance issues, duplication of responsibilities, unharmonised policies, lack of coordination and inadequate evaluation capacity generally hinder the policy performances of developing countries, which ultimately limit the MSP expectations. The benefits of MSP might not be readily achievable, since the MSP process itself takes time to establish after proper planning, implementation and adaptations. Therefore, the MSP all over the world will not be effective in an equal manner.

Promoting MSP in a large maritime domain for the first time might be challenging. Incomes, profits and savings, made possible through the MSP process shall be analysed to reduce the resistance that may initially arise from the stakeholders. The intellectual base of the MSP experts in the developing regions can be enhanced through capacity-building programmes, preferably with the cooperation of the European countries. Through a collaborative approach, the MSP can empower the IO countries, utilising Private-Public Partnerships for securing funds, infrastructure development and the essential know-how. Perhaps, regional programmes such as the Bay of Bengal LME programme can highlight the MSP in their action plans. An identification of the biological hotspots in the region, especially with regards to the trans-boundary LMEs occurring in the IO, is important. More research is needed to distinguish the Drivers, Pressures, States, Impacts and Responses (DPSIR) that determine the interactions between those ecosystems and society. Pilot programmes and stakeholder consultations are possible ways to raise the initial discussions on MSP, which can then be taken to the next level with strong political commitments.

REFERENCES AND NOTES

1. Rahmstorf, S., 'Thermohaline Ocean Circulation', *Encyclopedia of Quaternary Sciences*, Amsterdam: Elsevier, 2006.
2. Whisnant, R. and A. Reyes, 'Blue Economy for Business in East Asia: Towards an Integrated Understanding of Blue Economy', Partnerships in Environmental Management for the Seas of East Asia (PEMSEA), Philippines, 2015.
3. UN, 2016(a). Retrieved from Overall Assessment of Human Impact on the Oceans:http://www.un.org/depts/los/global_reporting/WOA_RPROC/Chapter_54.pdf
4. UN, 2016(b). Retrieved from The First Global Integrated Marine Assessment: http://www.un.org/depts/los/global_reporting/WOA_RPROC/WOACompilation.pdf
5. Dover, C., 'Impacts of Anthropogenic Disturbances at Deep-Sea Hydrothermal Vent Ecosystems: A Review', *Marine Environmental Research*, 102, 2014.
6. Zhang, Y., X. Su, F. Chen, Y. Wang, L. Jiao, H. Dong et al., 'Microbial Diversity in Cold Seep Sediments from the Northern China Sea', *Geoscience Frontiers*, 3, 2012.
7. Brenner, J., J.A. Jimenez, R. Sarda and A. Garola, 'An Assessment of the Non-Market Value of the Ecosystem Services provided by the Catalan Coastal Zone, Spain', *Ocean & Coastal Management*, 2010.

8. McNeely, J., 'People, Ecosystems and Climate: Governance Risks from the Degradation of Ecosysem Services in the Face of Ongoing Climate Change' *IRGC*, 2010.
9. Wafar, M., K. Venkataraman, B. Ingole, S. Ajmal Khan and P. Loka Bharathi, 'State of Knowledge of Coastal and Marine Biodiversity of Indian Ocean Countries', *PLoS ONE*, 6(1), e14613, 2011. http://doi.org/10.1371/journal.pone.0014613
10. Keesing, J., and T. Irvine, 'Coastal biodiversity in the Indian Ocean: The Known, Unknown and Unknowable', *Journal of Marine Science*, 2005.
11. Bhattathiri, P.M.A., 'Features at Some Significant Estuaries of India' in . *The Indian Ocean: A Perspective, Vol 1.* R. Sen Gupta and E. Desa (eds), New Delhi: Oxford and IBH Publishing Co.; 2001. pp. 271–297.
12. Wafar, M. et al., op.cit.
13. Benjamin, H., W. Shaun, S. Kimberly, K. Carrie, M. Fiorenza, A. Caterina, B. John, C. Kenneth, E. Colin, F. Helen, F. Rod, H. Dennis, L. Hunter, M. Elizabeth, P. Matthew, S. Elizabeth, S. Mark, S. Robert, W. Reg, 'A Global Map of Human Impact on Marine Ecosystems', Science, 319, 2008.
14. Freitas, B., L. Delagran, E. Griffin and K. Miller, K., *Too Few Fish: A Regional Assessment of the World's Fisheries*, Washington, DC: OCEANA, 2008.
15. De Young, C., 'Review of the State of World Marine Capture Fisheries Management: Indian Ocean', *FAO Fisheries*, 2006.
16. Martosubroto, P., *FAO Regional Review of the Eastern Indian Ocean*, FAO, 2005.
17. Rabotyagov, S., C. Kling, P. Gassman, N. Rabalais and R. Turner, R., *The Economics of Dead Zones: Linking Externalities from the Land to their Consequences in the Sea*', Center for Agricultural and Rural Development, 2012.
18. Miththapala, S., S. Sawarkar, P. Krishnan and M.M. Ariyabandu, *A Toolkit for Integrating Disaster Risk Reduction and Climate Change Adaptation in to Ecosystem Management of Coastal and Marine Areas in South Asia*, New Delhi: UNISDR/UNDP, 2012.
19. Sumaila, U., M. Samoilys, E. Allison, J. Cinner, C. De Yong and C. Kavanagh, *Economic Impacts of Ocean Acidification on Fisheries and Aquaculture in the Western Indian Ocean: Current Knowledge and Recommendations*, 2015.
20. Michel, D., H. Fuller and L. Dolan, 'Natural Resources in the Indian Ocean' in *Indian Ocean Rising: Maritime Security and Policy Challenges*, D. Michel and R. Sticklor (eds.),Washington: Stimson, 2012.
21. Bensch, A., M. Gianni, D. Gréboval, J. Sanders and A. Hjort, *Worldwide Review of Bottom Fisheries in the High Seas*, Rome: FAO, 2009.
22. Lal Mukherjee, A., 2013. Retrieved from Impact of tourism in coastal areas: Need of sustainable tourism strategy: http://www.marbef.org/wiki/Impact_of_tourism_in_coastal_areas:_Need_of_sustainable_tourism_strategy
23. Gall, S. and R. Thompson, 'The impacts of Debris on Marine Life', *Marine Pollution Bulletin*, 2015.
24. Wilson, S. and K. Verlis, 'The Ugly Face of Tourism: Marine Debris Pollution linked to Visitation in the Southern Great Barrier Reef Australia', *Marine Pollution Bulletin*, 2017.
25. Harriot, V., *Marine Tourism Impacts and their Management on the Great Barrier Reef*, James Cook University, 2002.
26. Roche, R., C. Harvey, J. Harvey, A. Kavanagh, M. McDonald, V. Stein-Rostang et al., 'Recreational Diving Impacts on Coral Reefs and the Adoption of Enviornmentally Responsible Practices within the SCUBA Diving Industry', *Environmental Management*, 2016.
27. Razak, I., K. Yusu, M. Salim and N. Salleh, 'Tourism Sea Activities that cause Damages

toward Coral Reefs in Samblian Islands', *Tourims, Leisure and Global Change*, 2014.

28. Cruz, R. V., H. Harasawa, M. Lal, S. Wu, Y. Anokhin, B. Punsalmaa et al., *Asia Climate Change, 2007, Impacts, Adaptation and Vulnerability*, 2007.
29. Solomon, S., D. Quin, M. Manning, M. Marquis, K. Averyt, M.M. Tignor et al., *Climate Change 2007: The Physical Science Basis*, Cambridge University Press, 2007.
30. Hoffman, J., *The Maldives and Rising Sea Level*, ICE, 2007.
31. Rahman, C., *Concepts of Maritime Security: A Strategic Perspective on Alternative Visions for Good Order and Security at Sea, with Policy Implications for New Zealand*, Victoria: University of Wellington, 2009.
32. Rumley, D. (ed.), *The Indian Ocean Region: Security, Stability and Sustainability in the 21st Century*, Australia India Institute, 2013.
33. Hardin, G., 'The Tragedy of the Commons', *Science*, 1968.
34. Creel, L., *Ripple Effect: Population and Coastal Regions*, Population Reference Bureau, 2003.
35. Cicin-Sain, B., R. Knecht, D. Jang and G. Fisk, *Integrated Coastal and Ocean Management: Concepts and Practices*, Washington DC: Island Press, 1998.
36. Douvere, F. and C. Ehler, C., 'Marine Spatial Planning: A Promising New Paradigm for applying an Antegrated, Ecosystem Approach to Ocean Management', Proceedings of the ICES 2008 Annual Science Conference, 2009.
37. *Chapter 17, Agenda 21.* Rio de Janeiro: UN, 1992.
38. Unger, S., A. Muller, J. Rochette, S. Schmidt, J. Shakeroff and G. Wright, *Achieving the Sustainable Development Goal for the Oceans*, Potsdam: Institute for Advanced Sustainability Studies, 2017.
39. Clarke, P. and S. Jupiter, *Principles and Practice of Ecosystem Based Management: A Guide for Conservation Practitioners in the Tropical Western Pacific*, New York: Wildlife Conservation Society, 2010.
40. Long, R., A. Charles, and R. Stephenson, R., 'Key Principles of Marine Ecosystem Based Management', *Marine Policy*, 57, 2015.
41. *UNESCO-IOC*, 2016(a). Retrieved from Marine Spatial Planning Initiative: www.unesco-ioc-marinesp.be/msp_faq
42. Pomeroy, R. and F. Douvere, F., 'The Engagement of Stakeholders in the Marine Spatial Planning Process, *Marine Policy*, 32, 2008, pp. 816–822. Retrieved from http://www.unesco-ioc-marinesp.be/uploads/documentenbank/a7c08c25e6fcea567c27ca564135f65a.pdf
43. Meiner, A., 'Integrated Maritime Poliy for the European Union: Consolidating Coastal and Marine Information to support Maritime Spatial Planning', *Journal of Coastal Conservation*, 14, 2010.
44. Rowder, L. and E. Norse, 'Essential Ecological Insights for Marine Ecosystem Based Management and Marine Spatial Planning', *Marine Policy*, 32, 2008.
45. Ehler, C. and F. Douvere, *Marine Spatial Planning: A Step-by-Step Approach toward Ecosystem-Based Management*, Intergovernmental Oceanographic Commission and Man and the Biosphere Programme, IOC Manual and Guides No. 53, ICAM Dossier No. 6. Paris: UNESCO, 2009.
46. Santos, C., T. Domingos, M. Ferreira, M. Orbach and F. Andrade, F., 'How Sustainable is Sustainable Marine Spatial Planning? Part I: Linking the Concept', *Marine Policy*, 49, 2014.
47. Ehler, C., *Marine Spatial Planning in the Arctic: A First Step Forward Ecosystem-Based Management*, Aspen Institute, 2011.
48. Ehler and Douvere, op.cit.
49. Day, J., 'Zoning: Lessons from the Great Barrier Reef Marine Park', *Ocean and Coastal Management*, 45, 2002.

50. Pitcher, R., P. Doherty, P. Arnold, J. Hooper and N. Gribble, *Seabed Biodiversity on the Continental Shelf of the Great Barrier Reef World Heritage Area*, CSIRO Marine and Atmospheric Research, 2007.
51. *UNESCO-IOC*, 2016 (b). Retrieved from Marine Spatial Planning Initiative: Great Barrier Reef Marine Park: http://www.unesco-ioc-marinesp.be/spatial_management_practice/australia_great_barrier_reef
52. Kenchington, R., and J. Day, 'Zoning, A Fundamental Cornerstone of Effective Marine Spatial Planning: Lessons learnt from the Great Barrier Reef Australia', *Journal of Coastal Conservation*, 15, 2011.
53. Kenchington, A., *Managing Marine Environments*, New York: Taylor & Francis, 1990.
54. Day, J., op.cit.
55. Ibid.
56. Rumley, D., and S. Chaturvedi, *Geopolitical Orientations, Regionalism and Security in the Indian Ocean*, South Asian Publishers, 2004.
57. Fatima, Q. and A. Jamshed, 'The Political and Economic Significance of Indian Ocean: An Analysis', *A Research Journal of South Asian Studies*, 30, 2015.
58. *IORA*, 2016. Retrieved from Blue Economy: www.iora.net/blue-economy/blue-economy.aspx
59. Smythe, T., 'Marine Spatial Planning as a Tool for Regional Ocean Governance? An Analysis of the New England Ocean Planning Network', *Ocean and Coastal Management* , 2017.
60. Vince, J., 'Ocean Governance and Marine Spatial Planning in Australia', *Australian Journal of Maritime and Ocean Affairs*, 6, 2014.
61. Ehler, C., *Coral Triangle Initiative: An introduction to marine spatial planning*. Jakarta, Indonesia: Coral Triangle Initiative on Coral Reefs, Fisheries and Food Security, 2013.
62. *SMSP*, 2015. Retrieved from Seychelles Marine Spatial Planning: On the leading edge of marine conservation and climate adaptation: www.cdb.int/doc/meetings/mar/soiom-2016-01/other/soiom
63. Allnutt, T., T. McClanahan, S. Andrefouet, M. Baker, E. Lagabrielle, C. McClennenet al., 'Comparison of Marine Spatial Planning Methods in Madagascar Demonstrates Value of Alternative Approaches', *PLoS One*,2012.
64. *CER*, 2016). Retrieved from Draft Marine Spatial Planning Framework: http://cer.org.za/wp-content/uploads/2016/08/Draft-Marine-Spatial-Planning-Framework-Published-for-public-comment-20160819-GGN-40219-00936.pdf
65. *ASCLME*, 2012). Retrieved from A strategic action programme for sustainable management of the Wester Indian Ocean Large Marine Ecosystems: http://asclme.org/reports2013/TDA%20supporting%20documents/10%20ASCLME_REPORT_PART_I_ver1_LR%20Lagabrielle.pdf
66. Pomeroy, R., K. Baldwin and P. McConney, 'Marine Spatial Planning in Asia and the Caribbean: Application and Implications for Fisheries and Marine Resources Management', *Desenvolvimento E Meio Ambiente*, 2014.
67. Michel, D., H. Fuller and L. Dolan, 'Natural Resources in the Indian Ocean' in, *Indian Ocean Rising: Maritime Security and Policy Challenges*, D. Michel and R. Sticklor (eds.), Washington: Stimson, 2012.
68. Pomeroy, R., op.cit.

7

Challenges of Developing the Ocean Economy from a SIDS Perspective

Romeela Mohee

Mauritius, a small island developing state in the Indian Ocean, presents several opportunities for developing an ocean economy. It has an EEZ of 2.2 million sq. km and an additional 4,00,000 sq. kms is shared with the island of Seychelles. One of the key components of the Government of Mauritius Vision 2030 is to develop the ocean economy by creating opportunities in the ocean-related sectors such as bunkering, port expansion, cruise tourism, seafood, fisheries and aquaculture for job creation and economic diversification. This chapter outlines the actions taken to promote the ocean economy in Mauritius and the challenges faced by the government. An analysis of the enabling factors, such as the institutional and legal framework, incentives for business facilitation and capacity-building efforts to boost the ocean economy are presented. The relevance of the ocean economy in international and regional agenda of the SDG 2030, Africa 2063 and in the SAMOA pathway is highlighted. The chapter concludes with an analysis of the priority sectors of the ocean economy for the next decade and the capacity-building initiatives in the formal educational sector.

Introduction

In 2013, the Government of Mauritius decided to adopt ocean economy as the next pillar of the economy. This was premised on the opportunities which were offered by its Exclusive Economic Zone of 2.2 million square km and an additional 4,00,000 sq. km which is shared with the island of Seychelles. Fig. 1 shows the Exclusive Economic Zone (EEZ) of Mauritius.

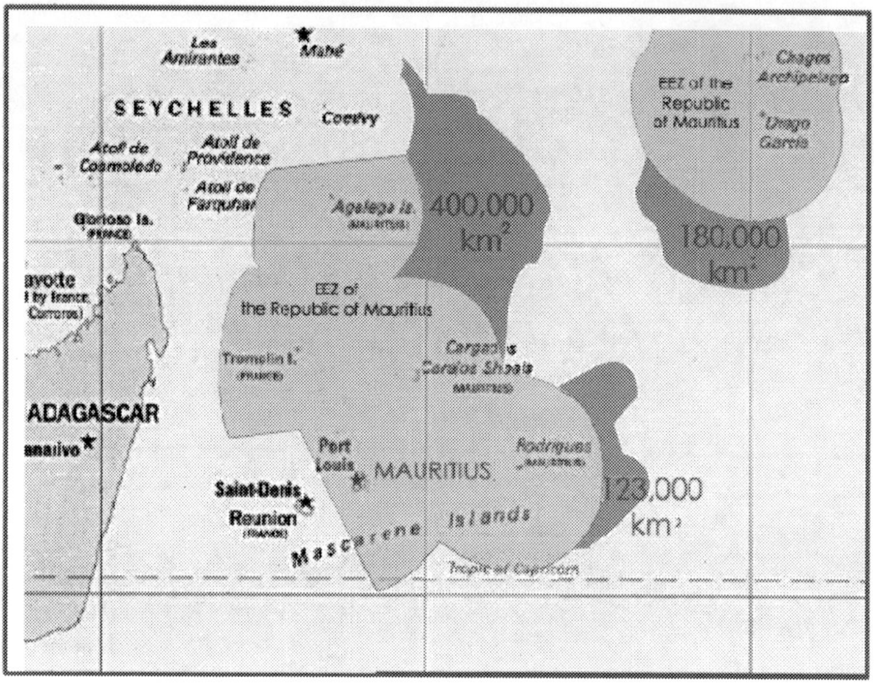

Figure 1. Exclusive Economic Zone of Mauritius[1]

The Ocean Economy Roadmap was launched in August 2013, during the National Dialogue on the Ocean Economy in Mauritius where the current status of ocean resources was presented and seven sectors were highlighted for development. Subsequently, an Ocean Task Force was set up under the aegis of the Prime Minister's office to drive developments/projects in the ocean sector. Following the launch of the National Dialogue and the Ocean Economy Roadmap, eight ocean economy related clusters were set up. Amongst these, one of them, the Ocean Knowledge Cluster, was chaired by the University of Mauritius and set up to identify actions to build capacity for supporting the ocean economy. The Ocean Economy Roadmap presented an analysis of the baseline and potential indicators, sector-wise, for the development of several maritime sectors as shown in Table 1.

Table 1. Key Indicators for Ocean-Related Sectors
(Baseline and short-term projections)[1]

	Sector	Baseline (2013)	Short-Term (2015)
A1	Bunkering and re-export of oil	280,000 Metric Tonnes	500,000 Metric Tonnes
A2	Fishing, seafood processing and aquaculture	Quantity (Tonnes): 178,500 Turnover (MUR BN): 15.75 Jobs: 12,150	Quantity (Tonnes): 241,000 Turnover (MUR BN): 17.65 Jobs: 14,000
A3	Coastal tourism and ocean-based leisure	Revenue (MUR BN): 1.26 Jobs: 2,800	Revenue (MUR BN): 1.84 Jobs: 3,000
B1	Renewable energy (Installed capacity)	N/A	Offshore wind: I MWDOWA cooling: 20 MW
B2	High value-added servicesand ship registry	Revenue (MUR BN): 381 Jobs: 240	Revenue (MUR BN): 582 Jobs: 500
B3	Prospection for hydrocarbonsand minerals (potential sites)	N/A	Oil and GAS: 0Minerals: 1
C1	Fivefold increase in output	GDP contribution (%): 11 Jobs: 18,000	GDP contribution (%): 14 Jobs: 21,000
C2	Centre of excellence for Ocean knowledge	No. of targeted research Projects: 20	No. of targeted research Projects: 30
C3	Mapping of living and non-living resources	Under 5% of maritime zone	7% of maritime zone
C4	Ecologically clean and safe EEZ	Identify baseline index	Improvement on baseline

In 2015, the new government took the strategic decision of creating a Ministry of Ocean Economy, Marine Resources, Fisheries, Shipping and Outer Island with a view to coordinate all the initiatives and projects in the ocean sector. The government's stated mission for the ocean economy was to set up a National Ocean Council (NOC) as the driving force of the Ministry and also to formulate a national ocean policy to develop the sector in the short, medium and long term. The main objective was to promote the development of economic activities and create jobs in a planned and economically sustainable manner. In the Presidential Address on 27 January 2015, the government's plan for 2015–2019 was outlined,[2] which stated that a "legal and a regulatory framework will be put in place to license, supervise, monitor and regulate the activities of the ocean related economic operators".

The International, Regional and Local Agenda

SDG 2030

The importance of the oceans and especially, protecting ocean-resources, has been well acknowledged in the Sustainable Development Goal 14 for the oceans which is dedicated to 'Life Below Water'. Goal 14 states, "Conserve and sustainably use the oceans, seas and marine resources". It urges all nations to develop knowledge on the ocean and related sectors and show progress towards meeting the SDG 14 targets of reducing pollution by 2025, to sustainably manage and protect marine ecosystems and promote sustainable practices as well as minimise the impact of ocean acidification, amongst others, by 2020.[3]

SAMOA Pathway

One of the outcomes of the 3rd International Conference on Small Island Developing States (SIDS) in 2014 included key recommendations[4] on Ocean Economy such as "Conserve by 2020, at least 10 per cent of coastal and marine areas of SIDS through protected areas and area-based conservation measures; address marine pollution; build resilience against invasive species; conservation; sustainable use/management of highly migratory fish stocks; protection of coral reefs and other vulnerable marine ecosystems; and scientific research and technical capacity-building".

Africa to Emerge as an Ocean/Blue Nation

The African Union initiation of "Harnessing the Blue Economy in Achieving the African Union Agenda 2063" is a firm step in Africa emerging as a blue nation. The African Union has pledged to develop the Blue Economy as one of its pillars and in 2015, it launched the '2015-2025 decade of African Seas and Oceans' and 'Africa Integrated Marine' (AIM) strategy. The seven missions embraced in the AIM 2050 strategy comprise several common elements with the Mauritian vision of the Ocean Economy, which are energy exploitation, climate change, environmental protection and conservation, safety of life and property at sea, maritime sector development including competitiveness, job creation, international trade, maritime infrastructure, transport, information, communication and technology and marine logistics.

Further, "in line with the aforementioned, the 2050 AIM strategy is hereby developed as a tool to address Africa's maritime challenges for sustainable development and competitiveness. The strategy aims to foster more wealth creation from Africa's oceans, seas and inland water ways by developing a thriving maritime economy and realizing the full potential of sea-based activities in an environmentally sustainable manner".[5]

These initiatives demonstrate that the Ocean/Blue Economy will be a strong determinant of the development of the African continent in the future.

Indian Ocean Region

Indian Ocean Rim Association (IORA)

The member states of the Indian Ocean Rim Association (IORA) – a grouping of 17 countries surrounding the Indian Ocean – signed a declaration[6] in September 2015, in Mauritius during the first IORA ministerial Blue Economy conference on enhancing Blue Economy cooperation for sustainable development in the Indian Ocean region. They agreed to focus on fisheries and aquaculture; renewable ocean energy; seaport and shipping and offshore hydrocarbons and seabed minerals. The objective was to promote the Blue Economy as a driver for sustainable development, encourage research and development and investment, technology transfer and capacity-building, so as to explore the full potential of the oceans and to ensure a sustainable development in the region.

Blue Economy of Seychelles

Seychelles was the first island nation in the Indian Ocean to adopt the Blue Economy. In 2014, in its vision for sustainable development in the Indian Ocean, the Minister of Foreign Affairs, Mr Alan, indicated that it was necessary to develop strategies to access the full potential of the ocean territory and called for a sustainable management of the seas. A blueprint[7] has been developed by Seychelles and several international/regional agencies are supporting Seychelles in this endeavour.

Ocean Economy of Mauritius

Similarly, in Mauritius, the development of the Ocean Economy is an emerging area with a very high potential and has a target of creating 25,000 jobs by 2030 in ocean-related sectors. The Ocean Economy has been identified as one of the key components of the Government Vision 2030,[8] which notes, "leveraging on the exclusive maritime economic zone to develop the ocean industry". One of the key pillars of the mission statement of the Prime Minister when he launched "Achieving the Second Economic Miracle And Vision 2030", in August 2015 was, "Port Louis harbour will be developed as a major port for the region so as to handle much higher traffic and other port-related activities. As regards the fishing sector, Government is negotiating with major international fishing companies for the setting up of fishing and seafood processing facilities locally".

The main sectors to achieve the Vision 2030 economic statement are maritime commerce, maritime security, marine energy, maritime education and scientific research (hydrography, oceanography, fisheries, coastal and inland training), ports and harbour management, promoting welfare of seafarers, environment management and coastal protection, maritime tourism (ashore and afloat), fisheries and aquaculture industry, shipbuilding and ship repair industries, maritime transport and auxiliary services, offshore exploration and exploitation, and maritime infrastructure development.

Actions Envisaged

As per the baseline indicators presented in Table 1, Mauritius was already exploiting some of the marine resources such as fisheries and the tourism sector. However, with an extended EEZ, an additional potential to develop the ocean-based industry has been created.

The National Ocean Council[9] set up seven working groups in August 2015 to devise the objectives, priorities and action plans for the following sectors:

(i) Fisheries and Aquaculture
(ii) Port and Shipping Development
(iii) Business Facilitation
(iv) Environment Protection Governance and Sustainable Development
(v) Capacity Building, Employment and Training
(vi) Seabed Exploration, Deep Ocean Water Applications (DOWA), Minerals, Renewable Energy, Marine Research
(vii) Regulatory Framework

These working groups have presented their action plans, including priorities and proposals, and provided an insight into the projects to be implemented. It was seen that some priority sectors would emerge as Mauritius would not be able to address the development of adequate human resources as also obtain the investment required to develop all the identified sectors. Therefore the sectors identified to be developed immediately were fisheries and aquaculture, shipping and ports development, seabed exploration, marine renewable energy and marine services and large-scale projects such as the DOWA project to provide an air conditioning of the public building from seawater. The key is to attract business, and investment from foreign players, to develop these sectors. Mauritius will therefore have to develop different economic models with overlapping elements by attracting investments from foreign and local companies in specific projects. A more long-term action plan also needs to be framed which is aimed at ensuring the sustainability of the ocean-related projects.

The DOWA Project

A private engineering company, Sotravic Ltd., has expressed interest in implementing the Deep Ocean Water Applications (DOWA) Project, which will consist of developing a seawater-based air conditioning system for public buildings within the capital city of Port Louis. This will be done in conjunction with another private company that will look into the offshore pipe alignment in the port area for the extraction of seawater.

Port and Bunkering

Port Louis harbour is the only port of Mauritius and is 1,000 km off the east coast of mainland Africa. It handles of 99 per cent of the country's external trade and contributes about 2 per cent of the GDP of the country. The expansion of the Mauritius Port Authority (MPA) includes extension and strengthening of the Mauritius container terminal berth and dredging works. For additional bunkering facilities, the MPA is allocating land in the port area for the construction of storage tanks. For bunkering, a private company 'BOMIN Ltd.' has recently started providing additional bunkering facilities on the island. There are plans to attract companies to set up more bunkering facilities.[10]

Seabed Exploration for Hydrocarbons and Minerals

The main activities of this sector fall under the aegis of the Prime Minister's office, the Continental Shelf and Maritime Zones Administration and Exploration Department and together with the Commonwealth Secretariat, a petroleum agreement (New Petroleum Bill), is being developed for Mauritius for developing the adequate legislation and appropriate regulatory instruments for offshore petroleum operation in Mauritius. The legislation for seabed exploration for minerals as well as for authorising research on the ocean resources is also being looked into.

Fisheries and Aquaculture

This is one of the main sectors identified to further develop economic activities in Mauritius. The Indian Ocean remains an important fish production area contributing about 23 per cent of the world's tuna production. Of this, 9,62,000 tonnes are produced in the Western Indian Ocean. In 2010, exports of fish and fish products comprising of tuna and fresh fish reached 1,04,740 tonnes (valued at USD 337 million), compared to 87,820 tonnes (valued at USD 300 million) for the previous year.[11]

The strategy of the Mauritius marine industry is based on sustainability, expansion of the industrial base, development of value-added fisheries and seafood and for harnessing a higher value in the related sectors including fishing, transshipment and ancillary services. Several measures have been taken to increase the production of fish stocks such as tuna and encouraging advanced artisanal fishing by training fishermen in using fish-aggregating devices when they venture out for fishing.

Seafood processing is being enhanced by fast tracking the setting up of commercial companies. New sites are being identified for aquaculture with emphasis on the involvement of the private sector. Existing setups like the Ferme Marine de Mahebourg are being expanded by having additional sites. Also, concession agreements are being developed for the lease of aquaculture sites. Businesses including Spirulina culture and Mollusc culture are also being planned.

Marine Renewable Energy

A marine renewable energy resource assessment and potential for its exploitation has been carried out by the Mauritius Research Council, with the support of the World Bank and the Commonwealth Secretariat. The first phase includes a decentralised micro grid and desalination system, to be set up in Rodrigues, a neighbouring island of Mauritius.

A collaborative agreement has been signed with Carnegie Wave Energy Ltd., Australia, to develop commercial wave energy applications using the CETO technology,[12] where buoys are installed at several levels and by the virtue of their movement, create a momentum for power generation. The feasibility of a kinetic water power plant, similar to the Rosch Technology, is also being studied.

Marine and Ship Registration

For Mauritius to develop into a hub, the Mauritius ship registry and its visibility is of the utmost importance. A ship registry of international standards which is compliant with international conventions such as the International Maritime Organisation (IMO) has to be developed. A campaign for involving global leaders to join this ship registry needs to be initiated. Assistance is also required to develop the appropriate regulating codes and for the computerisation of the registry.

Tourism Industry

Tourism has always been an important pillar of the Mauritian economy. The GDP linked to tourism was around 11 per cent in 2014.[13] Tourist arrivals were estimated to be around 1 million in 2016 and are expected to rise further.

It is planned to enhance the capacity of air carriers to develop the tourism industry and to further attract tourists from India and China. Studies have shown that Mauritius has been attracting high-spending tourists while promoting enclave tourism[14] and will continue to do so. Furthermore, the hotel development strategy[15] states that "priority will be given to hotel projects of the highest standard providing high quality service and belonging to 4 and 5 star categories".

Marine Biotechnology

Local universities and research institutes (Albion Fisheries Research Centre, Mauritius Oceanography Institute, Mauritius Research Council) are developing research projects for developing marine biotechnology-based industries, while consolidating existing work on corals and other marine organisms.

Future Projects

Future projects envisaged in the medium term include the development of a world-class aquarium, development of a fishing port quay, implementation of the masterplan of the Mauritius Port Authority, concession agreements for *barachois* development, relocation of "Afrintex" headquarters, recycling of the used ship oil and the development of additional facilities for a bunkering hub in Mauritius.

Several international experts from the World Bank, the Commonwealth Secretariat (UK), the Indian Ocean Rim Association (IORA), and the members of the National Ocean Council are working towards the development of the legal and regulatory framework of the NOC and for developing a National Ocean Policy.

Challenges Facing the Development of the Ocean Economy

For SIDS, oceans and seas constitute a much larger geographic area than their inland territory, when the EEZ is taken into account. Therefore, sustainable harvesting of marine resources presents a significant opportunity for environmentally sound, socially inclusive economic growth and development. However, there are a number of challenges facing the development of the Ocean Economy which have to be addressed.

Ensuring Environmental Sustainability

Like all SIDS, the Mauritian marine ecosystem is highly vulnerable to disasters. Coral reefs and seafood stocks in this region are particularly threatened by ocean acidification, overfishing, invasive species, and unsustainable marine tourism.

Ocean-related projects currently being identified in the Mauritian context are marine energy generation, seabed exploration, fisheries and aquaculture, expansion of port services, shipping, provision of bunkering facilities and cruise tourism. These projects will require activities such as exploration, drilling, dredging works, introduction of new species, installation of pipes in the ocean, and if unplanned, may lead to degradation and loss of marine habitat, disturbance to breeding grounds, marine pollution due to oil spills, release of toxics, and hazardous materials, air and noise pollution. Other environmental impacts include overexploitation of marine resources, conflicts between marine users, marine pollution and an overall environmental degradation of the marine resources. Hence, adequate measures have to be taken to mitigate any adverse effects related to the development and execution of these projects.

There is a need to focus on setting up a rigorous framework to ensure the environmental sustainability of the ocean resources. Mauritius being a SIDS, runs the risk of developing too many economic activities resulting in an overexploitation of the resources. The Ministry of Environment plays a key role by looking at the environmental impact assessment of all projects that are being set up by the authorities, ministries and the private sector before giving the green light and by streamlining environmental sustainability at all levels. A comprehensive exercise on Marine Spatial Planning is being undertaken with the World Bank by providing a coordinated planning and management of the seascape.

Lack of Data

The challenges of making informed marine management and policy decisions for sustainable development are the greatest for SIDS that are isolated in vast ocean areas. Here, disasters are often caused by ocean-related events such as tropical cyclones driven by warm ocean temperatures and cause damage to land and marine environment which endanger life and harm the economies of the SIDS.

To take evidence-based decisions and to develop rapid responses to natural disasters, knowledge about the ocean and reliable data is needed and the role of science and technology is crucial. Ocean science and monitoring allows us to better understand the ocean and climate system, and helps make better decisions for sustainable development.[16] The SIDS face long-term challenges in establishing, collecting and processing ocean observations. Limited resources are available to buy equipment and establish monitoring, and particularly the capability to repair expensive equipment is lacking.

Mauritius lacks the baseline data in several sectors and ocean monitoring, control and surveillance is important. Local scientists and engineers have a major responsibility in participating and contributing for developing the necessary local know-how, skills and competences in these emerging areas. At the Samoa Conference in 2014, the need for a greater engagement of scientists in policymaking, capacity-building and skills training in the ocean sector as well as the integration of local scientists into regional/international research teams and the development of strategic multi-sectoral partnerships was highlighted.

Financing of Ocean-Related Projects

Ocean-related projects require enormous infrastructure and sophisticated monitoring systems, which require large investments. The SIDS are dependent on wealthier economies/groups to help finance projects/improvements to uphold the international commitments. The developed countries have a responsibility to share new technologies and to develop a national capacity within the SIDS. These were highlighted during the Samoan Conference, with the 'Oceans' being one of the six priority multi-stakeholder partnership dialogues.

During the African Ministerial Conference[17] held on Ocean Economy and Climate Change in Mauritius in 2016, a special session was held on 'financing' in the presence of the African ministers and experts from the World Bank. The successful outcome was that international institutions would provide technical and financial assistance to support Mauritius in its ocean-related projects to address climate change challenges.

A business facilitation mechanism needs to be set up to encourage the private sector and foreign investors in developing ocean projects. Currently, the removal of financial regulatory barriers, such as for the exportation of tuna are being relooked into. A number of initiatives have also been taken by the Board of Investment (BOI) to promote business development with various activities such as:

(a) Creating a one-stop shop, single point of entry at the BOI, for ocean-related projects.
(b) Creating a repository of Ocean Economy projects.
(c) Creating a business facilitation desk at each ministry, as permits and licenses are normally granted at the level of different ministries/ departments.
(d) Streamlining of licensing processes using e-Licensing and e-Registration.

Institutional, Regulatory and Legal Framework

There are several government departments, institutions, agencies and private companies working in the ocean industry in Mauritius, as shown in Table 2. This results in an overlapping of activities and in some cases duplication amongst the institutions. For example, both the Albion Fisheries Research Centre and the Mauritius Oceanography Institute carry out research and development on coral farming. Hence, there is a need for clear demarcation of roles and responsibilities and proper coordination among the institutions to ensure effective delivery.

Table 2. Local institutions in Ocean-Related Sectors in Mauritius

Ministries	Private Firms	Training/ Research Institutions
Ministry of Ocean Economy, Marine Resources, Fisheries, Shipping and Outer Island[18]	Princes Tuna Ltd	Mauritius Oceanographic Institute
Ministry of Tourism and External Communications	IBL Ltd	Mauritius Research Council
Ministry of Environment, Sustainable Development, and Disaster and Beach Management	Ferme Marine de Mahebourg	Albion Fisheries Research Center
State Law Office, Attorney General's Office		Mauritius Maritime Training Agency (MMTA)
Prime Minister Office (Continental Shelf Department)		University of Mauritius- Faculty of Ocean Studies

There is also a lack of a proper legal framework. For example, in the Environment Protection Act, there is no mechanism in place to cater for the maritime zones. Also, the Maritime Zone Act does not provide any clear linkages on the institutional set up for protecting the maritime zone. No legislation governs the Intellectual Property Rights (IPRs) for joint research projects or authorisation for sampling/research of ocean resources. These gaps in relevant laws have to be addressed.

The setting up of the proper regulatory framework for the National Ocean Council (NOC) is being looked into. It was agreed by all parties that the NOC should provide a forum for both coordination and cooperation and will be the body to collect and share relevant data, to coordinate maritime surveillance and enforcement and also to engage with stakeholders. However, a public body can take different forms depending on the desired level of autonomy or the operational flexibility. Hence it has to be decided whether a department, an advisory body or an executive agency would be the most appropriate to match the chosen objectives/activities.

Lack of Human Resources and Capacity-Building

Trained and skilled people are required to develop, sustain and build the ocean economy-related activities. The job requirements identified in the working group on capacity-building under the aegis of the NOC, regrouping officers of the ministries, private companies and training institutions, concluded the following:

(a) *Cruise sector:* Currently around 1,100 people work in this sector. The Government programme aims at training 2,000 persons per year to work on cruise ships. Private companies indicate that around 14,000 vacancies would exist in this sector in the immediate future, with the visit of new cruise ships.

(b) *Shipping:* Jobs exist in this sector for seafarers and for technicians to work on-shore for repair of ships.

(c) *Fisheries and Aquaculture* (Oyster farms and Ferme Marine de Mahebourg expansion projects): Around 900 personnel will be needed in this sector.

(d) *Researchers/scientists and engineers:* Biologists, mechanical, civil engineers, marine environmentalists, technicians and marine veterinary personnel would be required.

Capacity Building to Develop the Ocean Economy

In 2013, a National Task Force led by the Prime Minister's Office was set up and the University of Mauritius (UoM) was designated to chair the Ocean Knowledge cluster. Subsequently, an advisory council consisting of public and private stakeholders was set up to analyse the ocean knowledge requirements and to advise on the way forward.

Two surveys were carried out by the University of Mauritius with the aim of compiling data on:

(a) The availability of trained human resources locally in the field of ocean studies, and
(b) On the training needs of other clusters under the National Task Force.

Based on the feedback obtained, a short, medium and long term plan was prepared for capacity building in the field of ocean studies.

It was revealed that amongst the tertiary education providers, only the University of Mauritius had formal programmes in marine sciences and related areas. In terms of need, it was seen that short-term technical and professional courses and programmes at the MSc level were needed for most of the areas while undergraduate courses would be needed more in the field of marine engineering and the shipping sector.

To address these issues, a Faculty of Ocean Studies was created at the University of Mauritius with three departments: Department of Marine and Ocean Science Fisheries and Mariculture; Department of Ocean Engineering and ICT and the Department of Maritime Trade and Finance.

In 2015, with the establishment of the NOC under the new Ministry of Ocean Economy, a working group on capacity-building, employment and training development was set up to identify the programmes required that would lead to job creation in the ocean-related activities and projects.

Skilled and trained people at the technical level, the engineering level and at the professional and middle management level as well as professional courses for seafarers and others would be required to drive these sectors. The Mauritius Maritime Training Academy (MMTA) has been set up and is actively running programmes for seafarers and for technical staff in the shipping industry to provide an adequate marine workforce at the vocational level including cruise activities and others.

The areas that require further training at professional and practitioner level are aquaculture, marine energy technicians, maritime law and trade, maritime business opportunities, general marine engineering and environmental protection of marine resources, prevention of dumping of toxic wastes, marine resources protection and climate change issues.

The impact of climate change on the ocean economy would be one of the key priorities which need to be addressed in Mauritius since being a SIDS, it is in a vulnerable situation and has a high threat from natural disasters. Also, the environmental sustainability of ocean related projects need to be emphasised.

Training Approach

Main instruments available for implementing training activities are listed as follows:

(a) Master courses: part-time programmes for officers already employed in public and private institutions.
(b) Undergraduate courses: marine resources management; marine sciences, marine energy, marine engineering, fisheries and aquaculture.
(c) Specialised technical courses for technicians to work in projects like the DOWA, renewable marine energy, ocean data management and for seafarers and technicians in the shipping sector.
(d) Certificate programmes with internships (to take advantage of the current GTES scheme set up by Government) leading to technical licensing for operators in the ocean economy.
(e) Short professional courses with government officers from Mauritius and African countries to enable then to implement and monitor ocean-related projects.
(f) Short technical training courses: highly focused summer schools, workshops, webinars, e-learning modules at the certificate level.
(g) Developing analytical tools for ocean data management and marine spatial planning.
(h) Undertaking research programmes, with an applied and developmental perspective in the various ocean economy sectors.

Conclusion

The ongoing and future projects carried out by the various actors in the promotion of ocean activities at the national level would, in the long term, help in the development of new sectors, and will result in an increased energy security, food security, increased investment in ocean projects and in the creation of knowledge and sustainable partnerships.

There is immense potential in developing training facilities to address the human resources requirements for the development of the Ocean Economy. This would require identifying sectors with a high growth potential (such as ports, aquaculture, renewable energy, tourism and marine biotechnology) which could further support economic development and sustainable job creation. Investments in capacity-building will be crucial if Mauritius is to ensure a sustainable development of its coastal and ocean resources. Also, the pursuit of scientific research in the related multidisciplinary areas of ocean economy will be crucial to the creation and transfer of knowledge that will lead to the success of this pillar of the Mauritian economy.

REFERENCE AND NOTES

1. 'Ocean Economy Road Map 2013', January 2014. Retrieved at http://www.oceaneconomy.mu/PDF/Brochure.pdf.
2. 'Government Programme 2015-2019'. Retrieved at www.lexpress.mu/ites/lexpress/.....govprog2015.pdf
3. 'Sustainable Development Goal 14', January 2017. Retrieved from https://sustainabledevelopment.un.org/sdg14
4. 'SAMOA Pathway Outcome; December 2016. Retrieved at www.sids2014.org/index.php?menu=1537
5. 'AIM 2050 Strategy', August 2016. Retrieved at pages.au.int/maritime/documents/2050-aim-strategy).
6. 'IORA Blue Economy Ministerial Conference', August 2016. Retrieved at www.iora.net/media/158070/mauritius_blue_economy_declaration.pdf
7. 'Seychelles Blue Economy Paper', December 2016. Retrieved at thecommonwealth.org/.../seychelles-blue-economy-strategic-roadmap-and-implementation
8. 'Vision 2030', September 2016. Retrieved at www.govmu.org/Achieving the second economic miracle and vision 2030.pdf.
9. 'National Ocean Council', Ministry of Ocean Economy, Marine Resources, Fisheries, Shipping and Outer Island, August 2015. Retrieved at http://fisheries.govmu.org/English/Pages/default.aspx
10. 'A Bunkering Hub: Driving the Ocean Economy' December 2016. Retrieved at http://www.globalfinance.mu/index.php?option=com_content&view=article&id=537:mauritius-a-bunker-hub-driving-the-ocean-economy&catid=15&Itemid=963
11. 'Board of Investment', December 2015. Retrieved at http://www.investmauritius.com/investment-opportunities/ocean-economy.aspx
12. 'Marine Renewable Energy for Mauritius', January 2017. Retrieved at http://www.un.org/Depts/los/consultative_process/icp13_presentations-abstracts/2012_icp_abstract_suddhoo.pdf.
13. World Travel and Tourism Council, 'Economic Impact 2015', January 2017. Retrieved at www.wttc.org/media/files/reports/economic impact research

14. Nunkoo and Ramkisoon, 'Stakeholders' Views of Enclave Tourism: A Grounded Theory Approach', *Journal of Hospitality and Tourism Research.*
15. 'Hotel Development Strategy (HDS)', Ministry of Tourism, November 2016. Retrieved at www.gov.mu
16. Grimes. S. 'Ocean Data and Figures'. *Scidev.* Retrieved at www.scidev.net/.../ocean-science-development-sids-facts-figures.htmlý
17. African Ministerial Conference on Ocean Economies and Climate Change Mauritius Communiqué as agreed at "Towards COP22", September 2016. Retrieved at www.worldbank.org/.../mauritius-communique-as-agreed-at-towards-cop22-african.
18. Ministry of Ocean Economy, Marine Resources, Fisheries, Shipping and Outer Island, January 2015. Retrieved at www.govmu.org.

8

Implementing the Blue Economy in the Western Indian Ocean: Marine Protected Areas and Integrated Coastal Management in Seychelles

Nirmal Jivan Shah

The recent and dynamic pursuit of the Blue Economy globally has been led by Small Island Developing States (SIDS) and in the Indian Ocean particularly by the Republic of Seychelles. The Blue Economy concept as conceived by Seychelles moves away from the business-usual approach to marine management and introduces strong sustainability elements. As such, environmental protection, which Seychelles has always been famous for, is one of the key pillars. Seychelles is taking its historical strengths in Integrated Coastal Management (ICM) and in Marine Protected Areas (MPAs) to build a new vision and implementation modality for the Blue Economy. This chapter highlights several new and innovative MPA and ICM initiatives emanating from both Government and civil society that are lynchpins to a new governance paradigm for the Seychelles' Exclusive Economic Zone (EEZ) and are also a model for other countries. These include a Marine Spatial Plan (MSP) for the entire EEZ, a new and innovative funding mechanism for the implementation of the MSP, and major advances in coral reef restoration in the MPAs.

Introduction

The Blue Economy is being heralded as an innovative agenda for setting the pace towards a new governance of the ocean.[1] As a new development concept, the Blue Economy needs unpacking. "We need to know what's under its hood, how it works and how to fuel it. More than that, we need to know what it is not - the Blue Economy cannot be new branding for business as usual or old wine in a new bottle".[2] A number of disciplines from the natural to social sciences must be involved for the Blue Economy to be truly game-changing; it must be multidisciplinary, multi-sectoral and multi-partner and it must respond to societal needs rather than only the needs of the commercial markets. Therefore, it must involve a number of actors and stakeholders working together and across governmental, private and civil society sectors.[3] The Small Island Developing States (SIDS), sometimes characterised as Large Ocean States because they have small land masses and populations but very large marine Exclusive Economic Zones (EEZs), are particularly challenged in this regard.[4]

The Unique Challenges of Seychelles as a Small Island Developing State

The Republic of Seychelles is an archipelago consisting of 115 islands, with a total land area of 455 sq. km, and an EEZ of 1.37 million sq. km.[5] The Seychelles Bank (4-5°S, 54-57°E) is a shallow (less than 100m water depth) and low-lying 'plateau' (2–6 m above sea level on average). The Bank is surrounded by the deep Indian Ocean, reaching depths of thousands of metres. There are two groups of islands - 43 high granitic islands and 72 lowcoralline islands. The granitic islands are the only oceanic islands in the world made up of continental granite and are of considerable scientific importance having been isolated from continental land masses for over 70 million years resulting in a unique biodiversity.[6] The coastline length is estimated at 746 km and the area of the continental shelf at 31,479 sq. km. The population was 93,400 in 2015. The Seychelles society is ethnically mixed resulting in a heterogeneous society with no traditional community structures.

In 2015, Seychelles was classified as a high income country by the World Bank, with a Gross Domestic Product (GDP) per capita of Seychelles Rupees 194,889 (over US$ 15,000). The real GDP growth rate was 4.3 percent in 2015. Seychelles has managed to fully achieve most of the Millennium Development Goals (MDGs) with some targets such as health, education and environment having been attained well before 2000.[7] However, the country faces a number of economic challenges common to the SIDS, such as remote location, lack of natural resources, small economy and a very small population.

Climate change and the related projected increases in sea surface temperatures and sea-level rise threaten the socio-economic development in Seychelles. Vulnerabilities are mostly in fisheries, agriculture, water resources, human settlements and human health. Coral bleaching, caused by warm ocean temperatures, threatens the coral reefs that provide valuable ecosystem services for the people and ecological functions for hundreds of marine species.[8] The ongoing petroleum exploration must also be seen in the context of possible anthropogenic impacts especially if the oil is discovered.[9]

The Economic Basis for the Blue Economy

The two key pillars of the Seychelles economy are tourism and fisheries, thus easily forming the foundation of a Blue Economy. The shallow banks of the Seychelles provide vital food security and employment through a small but very important artisanal fishery. This fishery is operated exclusively by the Seychellois and uses several types of boats and fishing gear.[10] As a fish-dependent country, Seychelles relies heavily on it to provide essential food and nutrition security to the local population. Some key fish species such as Jacks (*Carangidae*) form the bulk of the landings but up to 100 species can be caught. The fish landing in 2015 was 3,348 tonnes including sea cucumbers, lobster and octopus. The oceanographic conditions around Seychelles provide excellent habitat for large stocks of tuna. Given the migratory nature of tuna in this part of the ocean, the pattern of catch is spread throughout the year across the Seychelles EEZ. Seychelles has been serving as the main regional hub for the industrial purse seine fleet in the Western Indian Ocean (WIO). The vast majority of the frozen purse seine catch in the WIO, about 80 per cent, is either landed for processing/canning in Seychelles (30 percent) or transshipped for processing elsewhere in the Western Indian Ocean (70 per cent).[11] In 2015, 32,068 tonnes of tuna were canned in Seychelles.

The spectacular and pristine landscapes and seascapes have enabled Seychelles an easy entry into tourism. The country's tourism policy has always been environmentally friendly and eco-tourism was seen as a viable alternative as far back as the 1970s[12]. While Seychelles has always been seen as a sun and sea destination, the current policy on tourism development continues with the objective of positioning Seychelles as an exclusive and quality destination, setting it apart from others. It advocates coordinated national efforts to maintain and expand the tourism sector's contribution to the economy in a manner that is environmentally and socially sustainable. It targets eco-tourism development under its Eco-Tourism Strategy.[13] However, the country eschews mass tourism in favour of a small number of select, discerning and high-value clients. In 2016, 3,02,000 visitors landed in Seychelles – the largest number ever recorded. The top nationalities were European, numbering almost 1,90,000 with France, Germany, Italy and the United Kingdom in the top four respectively.[14]

Transiting to the Blue Economy

Over a decade ago, the Seychelles government started to push the regional and global agenda towards adopting the Blue Economy as a viable development paradigm. This makes perfect sense because as a SIDS, sitting on small pieces of land, seemingly "sea-locked" within large a large ocean territory, the country has struggled with various land-based development paradigms such as the Green Economy.[15] The Blue Economy concept encompasses fisheries and tourism sectors, as well as the exploration of undersea resources and minerals, ocean-based renewable sources of energy and transportation through its sea routes. The government believes it is inappropriate to purely re-label the current uses of marine resources as "blue". It believes that the requirement for sustainability is predicated on a paradigm shift in both the mind-set of those currently exploiting the ocean resources and the diversity of ways in which the marine environment is used for the benefit of humankind.[16]

At the core of the Seychelles' Blue Economy concept is the de-coupling of socio-economic development from environmental degradation. To achieve this, the Blue Economy approach is being founded on a strong foundation of environmental protection and the further assessment, incorporation and implementation of the real value of the natural (blue) capital into all aspects of economic development.[17] The Government has set up the Blue Economy Department in the Ministry of Finance, Trade and Blue Economy and more recently moved it to a higher level in the Vice President's office. However, few others outside the government have managed to fully grasp the new challenge, and the NGO Nature Seychelles is the only civil society organisation that has set up a structure – The Blue Economy Knowledge Centre to build knowledge, awareness and projects on the subject.

Marine Protected Areas and Integrated Coastal Management (ICM) in Seychelles

Seychelles has a long tradition of successful environmental protection,[18] which has been led by Government with the help of international organisations and more recently, by NGOs. The current legal basis for environmental protection is enshrined in the Constitution of Seychelles. Article 38 notes "The State recognizes the right of every person to live in, and enjoy a clean, healthy and ecologically balanced environment and with a view to ensuring the effective realization of this right, the State undertakes:

(a) To make measures to promote the protection and preservation and improvement of the environment;
(b) To ensure a sustainable socio-economic development of Seychelles by a judicious use and management of the resources of Seychelles; and
(c) To promote public awareness of the need to protect, preserve and improve the environment."

The country has implemented two Environmental Management Plans since 1990 and its first National Biodiversity Strategy and Action Plan (NBSAP) in 1997. Currently, the Seychelles Sustainable Development Strategy (SSDS), 2012-2020 and the NBSAP 2015-2020 are on hand to guide development. It has some of the oldest marine and coastal protected areas in the WIO, and about 50 per cent of its land area is protected. Further, it has the most extensive network of Marine Protected Areas (MPAs) of any country in the WIO, comprising 15 sites of various designations and management regimes and covering an estimated 230 sq. km of ocean. As regards the Integrated Coastal Management (ICM), the broadest regional approach in the Western Indian Ocean region was started in 1993 at the Arusha Conference and spread throughout the region including Seychelles.[19] Seychelles has been an acknowledged as a leader in ICM.[20] In 1994, the ICM was enshrined as a legal principle in the Environmental Protection Act of Seychelles (EPA 1994) and remains so in the new EPA 2015. The country has undertaken many ICM initiatives and it has been incorporated in land use planning as well.

Advances in MPAs and ICM within the Blue Economy Framework

Three innovative and ground-breaking initiatives to push the boundaries of ICM and that of MPA design and management to meet challenges particularly to do with climate change that could aid progress towards the Blue Economy are described below. Two of these are government projects and one by the civil society.

Growing MPAs Beyond International Targets: The Marine Spatial Plan

The Government has recently taken bold new commitments to expand the Protected Areas estate beyond the Aichi Targets set by the CBD Strategic Plan for Biodiversity 2011–2020. Seychelles has committed to up to 30 per cent of its Exclusive Economic Zone (covering 4,00,000 sq. km) as Marine Protected Areas through a Marine Spatial Plan. This is to ensure that the species and habitats have a long-term protection, to improve the resiliency of ecosystems faced with climate change and to ensure economic opportunities for fisheries, tourism and other uses. The 30 per cent to be protected will be zoned in high and medium biodiversity protection areas – each about 15 per cent of the EEZ – and compatible uses will be identified to align with the biodiversity objectives. The Ministry of Environment, Energy and Climate Change is the implementing agency and The Nature Conservancy is providing technical assistance. A Steering Committee made up of representatives from the government, private and civil society sectors is guiding the process. The project is funded by the Government of Seychelles, private grants to The Nature Conservancy and grants from GoS-UNDP-GEF. The MSP Initiative is a part of the government-led 'Debt for Adaptation Swap' that has created the Seychelles Conservation & Climate Adaptation Trust – SeyCCAT. Phase 1 of the Plan was launched in 2014 and lays the foundation for a coarse-scale zoning design and draft management considerations which were to conclude in late 2016. Phase 2 will include refining the zoning design and management considerations, developing an implementation plan and new regulations - Phase 2 will conclude in 2020.[21] This is the first comprehensive Marine Spatial Plan in the WIO. Once approved, it will be the second largest in the world.

Funding Management of the New MPAs: Debt for Adaptation Swap

The country has started a unique Debt for Adaptation Swap through the creation of the Seychelles Conservation & Climate Adaptation Trust (SeyCCAT). It links a restructuring of some of the island nation's international debt with a financial mechanism to support adaptation to climate change, namely through an improved marine and coastal ecosystem management. That management will include the Marine Spatial Plan mentioned above. Seychelles is the first country in the world to set up a funding mechanism for its Marine Spatial Plan before the plan is completed. The debt restructuring, facilitated by The Nature Conservancy (TNC) with the involvement of the Paris Club of creditor nations and South Africa, includes using private investment to buy back US$ 30 million of Seychelles debt at a discounted rate.

In turn, the Seychelles government agrees to the following:

(i) Complete a marine spatial plan; create the SeyCCAT to fund implementation of the plan as well as other conservation and climate adaptation projects; and
(ii) Pay back the private investors over 20 years.[22]

The climate adaptation elements will involve coastal defence strategies, coral reef restoration and other mechanisms. These will be mostly in the form of competitive grants to relevant national organisations. Seychelles has successfully concluded the financial part of the debt buyback process with the Paris Club. The buyback initiative is a success thanks to a concessionary loan from the TNC and generous grants from a group of international marine conservation foundations. The SeyCCAT has been set up by law and trustees met throughout 2016. A Chief Executive for the Trust is being recruited so that the implementation plan can be rolled out in 2017.

Adapting the MPAs to Climate Change: Geo-Engineering Coral Reefs

Coral reefs provide a wealth of ecosystem services, including food, recreational use, biodiversity benefits, and regulating services, that are vital to the local economies and food security of human populations living in vulnerable coastal areas.[23] However, coral reefs are probably the ecosystems which are most at risk in Seychelles as a result of a range of direct human activities and climate change. In 1997-1998, an El-Nino coupled with an Indian Ocean Dipole event resulted in the highest seawater temperature anomalies recorded in 50 years and a worldwide mass coral bleaching event.[24] The reefs in the WIO region were the most severely impacted: coral mortality was 30 per cent at the regional level with 80–95 per cent on the badly affected reefs in the Seychelles, and live coral cover reduced to less than 3 per cent in some areas.[25] While some reefs recovered naturally within 5–10 years, most others remained as rubble-strewn wastelands even within well-established MPAs.[26]

In order to ensure that the reefs continue to provide the local population with all the vital biological, ecological and socio-economic goods and services, the USAID Development Grant Program (DGP) and the GEF through

UNDP and the Government of Seychelles funded a geo-engineering project by the NGO Nature Seychelles from 2011–2015 to initiate the Indian Ocean's first large scale coral reef restoration at a degraded reef site within the Marine Protected Area of the Cousin Island Special Reserve. The project used the reef gardening concept, by generating a pool of farmed colonies in underwater nurseries until they reached a threshold transplantation size. This was followed by transplanting the nursery colonies onto the degraded reefs. The project built and cultivated 12 mid-water nurseries with more than 40,000 coral fragments of 32 coral species. The total area transplanted was 5,225 sq. m, which is thought to be the largest coral reef restoration project in the world using this particular method and considered as a leading "Blue Solution" for Africa,[27] as well as an innovation.[28] The project has attracted attention beyond the conservation and scientific domains, and has been featured in international media, including CNN.[29] Another smaller reef was planted over an area of about 1,400 sq. m. In 2016, a severe El-Nino caused widespread bleaching across the Seychelles.[30] Whist the Nature Seychelles' "designer reef" resisted bleaching for the first 3 months, most of its corals bleached similarly at other sites. However, "super corals" that have not bleached at all have been discovered by Nature Seychelles, and 3,000 fragments of these are already being grown in nurseries for the second generation of designer reef engineering.

Conclusion

The Western Indian Ocean supports some 60 million people living within 100 km of the shore. The annual economic output of this region is estimated to be the fourth-largest after the region's biggest economies – South Africa, Kenya and Tanzania. It produces more than US$ 20.8 billion in goods and services every year.[31] Most of the countries of the region have initiated plans for the Blue Economy or Ocean Economy.[32] However, the smallest country in the region, the Republic of Seychelles, has shown itself as the leader in propelling a new vision for the Blue Economy and not simply re-branding the business-as-usual policies and practices. Seychelles, with its long history of successful environment protection and its recent graduation to a high-income country status, is nevertheless navigating stormy waters. As a tiny nation of less than 1,00,000 people, with 115 islands scattered over a large EEZ, with no resources such as oil and minerals and far from knowledge and market centres, it doesn't obviously lend itself to optimistic scenarios. But, some knowledge and implementation partnerships between the government and civil society organisations have shown early signs that inadequate human and financial capacity can be leapfrogged through synergy and innovation.[33] The Blue Economy offers a platform, more than any other, for a robust sustainable development involving all actors.

REFERENCES AND NOTES

1. Michel J.A., *Rethinking the Oceans – Towards the Blue Economy*, US: Paragon House, 2016.
2. Shah, N.J., 'Blue Economy needs R&D', 2013. Retrieved from http://theenvironmentinseychelles.blogspot.com/2013/11/the-blue-economy-wheres-research-and.html Accessed January 13 2017.
3. Nature Seychelles, 'Blue Economy: Notes for Blue Printing'. Retrieved from http://www.natureseychelles.org/knowledge-centre/scientific-papers-database/blue-economy-seychelles-1/221-blue-economy-blueprint-notes/file Accessed January 13 2017.
4. UNCTAD, 'The Oceans Economy: Opportunities and Challenges for Small Island Developing States', UNCTAD/DITC/TED/2014/5, 2016.
5. Shah, N.J., 'Coastal Zone Management in Seychelles' in Proceedings of the National Workshop on Integrated Coastal Zone Management in Seychelles, Government of Seychelles, World Bank and Sida, 1995.
6. Shah, N.J., R.A. Payet and K. Henri, 'National Biodiversity Strategy and Action Plan of Seychelles' Government of Seychelles
7. Ministry of Foreign Affairs, 'Seychelles Millennium Development Goals Status Report 2015', Government of Seychelles/UNDP, 2015.
8. Government of Seychelles, 'Intended Nationally Determined Contribution (INDC) Under The United Nations Framework Convention On Climate Change (UNFCCC)', Republic of Seychelles, 2015. Retrieved from http://www4.unfccc.int/submissions/INDC/Published%20Documents/Seychelles/1/INDC%20of%20Seychelles.pdf Accessed on 12 January 2017.
9. Shah, N.J. and P. S. Plummer, 'Towards the Development of Ecologically Sustainable Petroleum Industries in Island States: The Seychelles Example', Proceedings of the First Indian Ocean Petroleum Seminar, UNDTCD, 1992.
10. Lablache. G., G. Augier de Moussac and N.J. Shah, 'Summary Description of the Artisanal Fisheries and Resources for the Seychelles', FAO, 1988. Retrieved from http://www.fao.org/docrep/field/303859/3038590a.htm Accessed on January 10 2017
11. POSEIDON, MRAG, NFDS and COFREPECHE, 'Review of Tuna Fisheries in the Western Indian Ocean' MARE/2011/0, Brussels: European Union, 2014.
12. Shah, N.J., 'Nature and Tourism in Seychelles' in *Tourism of Quality: A Concept of Tourism Development Compatible at the Same Time to Economic, Social and Environmental Aspects*, Switzerland: AIEST Press, 1991.
13. Government of Seychelles, 'Vision 21: Tourism Development in Seychelles', 2001. Retrieved from http://www.natureseychelles.org/knowledge-centre/scientific-papers-database/national-documents/151-vision-21-tourism-development-in-seychelles-2001-2010-2001/file. Accessed on January 3, 2017.
14. National Bureau of Statistics, 'Seychelles in Figures', Government of Seychelles, 2016.
15. Shah, N.J., 'Our Ocean Moment', 2016. Retrieved from http://www.blueeconomyseychelles.org/34-our-ocean-moment. Accessed on January 12, 2017.

16. SRSL, 'Developing a Blue Economy Roadmap in The Republic of Seychelles: Contribution of New and Emerging Marine Sectors', Government of Seychelles/Commonwealth Secretariat, 2015.
17. Government of Seychelles, 'Seychelles' Blue Economy Roadmap: Defining a Pathway to Prosperity. Ministry of Finance, Trade and the Blue Economy, 2015.
18. Shah, K. J., 'Nature Conservation in Seychelles' in Proceedings of the Symposium on the Indian Ocean and Adjacent Seas: Their Origin, Science and Resources, *J. mar.biol. Ass. Ind.*, 1972, 14 (1): 312-314.
19. Shah, N.J., O. Linden, C. G. Lundin and R. Johnstone, 'Coastal Management in Eastern Africa: Status and Future', *Ambio*, 26: 4, 1997, pp. 227-234.
20. Shah N.J., 'Managing Coastal Areas in the Seychelles', *Nature & Resources*, 31(4), 1995, pp. 16-33.
21. Anon, 'Seychelles Marine Spatial Planning Initiative 2016', 2016. Retrieved from http://www.seychellesmarinespatialplanning.com/ 4_v4_17Mar2016.pdf Accessed on January 6, 2017.
22. Blue Economy Seychelles, 'Debt Swap to boost Conservation, 2016. Retrieved from http://www.blueeconomyseychelles.org/debt-swap-to-boost-conservation. Accessed January 15, 2017.
23. Sheppard. C., S.M. Wells and M.D. Jenkins, *Coral Reefs of the World, Vol. 2: Indian Ocean, Red Sea and Gulf,* Gland: IUCN, Conservation Monitoring Centre, 1988.
24. Spencer, T., K. Teleki, C. Bradshaw, and M. D. Spalding, 'Coral Bleaching in the Southern Seychelles during the 1997–1998 Indian Ocean Warm Event', *Marine Pollution Bulletin,* 40(7), 2000, pp. 569–586.
25. Obura, D., 'Resilience and Climate Change: Lessons from Coral Reefs and Bleaching in Western Indian Ocean', *Estuarine, Coastal and Shelf Science,* 63, 2005, pp. 353–372.
26. Spalding. M.D. and G.E. Jarvis, 'The Impact of the 1998 Coral Mortality on Reef Fish Communities in the Seychelles', *Marine Pollution Bulletin,* 44, 2002, pp. 309–321.
27. Blue Economy Seychelles, 'Seychelles Project chosen as an African Blue Solution' 2016. Retrieved from http://www.blueeconomyseychelles.org/item/52-seychelles-project-chosen-as-an-african-blue-solution.Accessed on January 12 2017.
28. Estico, X., 'Establishment of Local and Regional Innovation Ecosystem for Reef Restoration: Proceedings of the International Workshop on Establishing The Regional Innovation Platform based on STP' UNESCO-WTA, 2015.
29. CNN, 'Inside Africa, Reef Rescuers Race against Time', Video, 2015. http://edition.cnn.com/videos/intl_tv-shows/2015/06/29/seychelles-biodiversity-inside-africa-b-spc.cnn/video/playlists/intl-inside-africa
30. Blue Economy Seychelles, 'Status of Coral Reefs after the Recent Bleaching Event, 2016. Retrieved from http://www.blueeconomyseychelles.org/item/53-the-status-of-coral-reefs-after-the-recent-bleaching-event. Accessed on January 13, 2017.
31. Obura, D, 'Reviving the Western Indian Ocean Economy: Actions for a Sustainable Future', Gland: WWF International, 2017.
32. Economic Commission for Africa, 'Africa's Blue Economy: A Policy Handbook', Addis Ababa: UNECA, 2016.
33. Shah, N.J., 'We are GODS: Blue Economy and Great Ocean Developing States', 2014. Retrieved from https://blueeconomyseychelles.wordpress.com/2014/09/ Accessed on January 12, 2017.

9

Maritime Security as an Enabler of the Blue Economy

S. Parmesh

Ocean resources have served humankind for centuries and the entire life support system of our planet is dependent on the oceans, since they have the potential to absorb and regenerate. However, a growing population, shrinking land resources, unprecedented contraction of global output and human-induced degradation of environment have caused a deleterious effect on the oceans. The concept of the Blue Economy advocates the undertaking of regenerative and restorative ocean activities to ensure sustainable oceans and enhanced economic development through blue industries. The Blue Economy, however, is also fraught with challenges that require good ocean order and rule of law. A robust maritime security mechanism is thus an integral part of the sustenance of the Blue Economy since no development/business activity can thrive in an atmosphere of disorder/insecurity.

Introduction

Ocean resources have served humankind for centuries and have been a significant source of growth and sustainable economic development. The contribution of the maritime medium to the global trade and commerce and as a significant source of food and energy has been immense.[1] In fact, the entire life support system of our planet is totally dependent on the oceans, since they have the potential to absorb and regenerate. Today, the economic relationship with the oceans is once again evolving in newer ways and the oceans are becoming a focal point in the discourse on growth and sustainable development.

The coastal states depend on oceans for fishing, renewable energy, transport, minerals, oil and gas, other living and non-living resources for earning livelihood and socio-economic growth of its people. In almost 54 coastal and island countries, up to two-thirds of the total national territory is comprised of the ocean. Further, for billions around the world – especially the world's poorest – healthy oceans mean jobs, food and protection. The FAO estimates that fisheries and aquaculture assure livelihood for 10–12 per cent of the world's population with more than 90 per cent of those employed to capture fisheries working in small-scale operations in developing countries. It is also estimated that coastal areas within 100 kilometres of the ocean, account for about 61 per cent of the world's total Gross National Product (GNP) and therefore are of particular importance for the developing countries.[2] However, growing populations, shrinking land resources, unprecedented contraction of global output, and human-induced degradation of environment have induced the effect of climate change on the planet.

Further, the rising ocean temperature and atmospheric carbon have resulted in ocean acidification which has impacted negatively on their balance and productivity. It is imperative that the efficiency and optimisation of natural resources in the oceans are sustained to maintain the vitality of not only the oceans but also the entire life support system of the planet.

Concept of Blue Economy

The concept of Blue Economy is the decoupling of socio-economic development from environment degradation, i.e. it advocates undertaking regenerative and restorative ocean activities that lead to enhanced human health and well-being including ensuring food security and sustainable livelihood. Following the Rio+20 Declaration in 2012, the participating countries came together to advance the concept of 'Green Economy' where sustainable development and poverty eradication were the key pointers. The island countries, however, argued that oceans also need to be given their due. Coastal and island states promoted Blue Economy by advocating that the oceans had a major role in the sustenance of humanity and that the Blue Economy was a better development approach that suited their circumstances.

The Blue Economy Paradigm and India

As India looks to the oceans in search of energy and food security, it is prudent that its policy decisions are dictated by the ill-effects of sea encroachments, water pollution, dumping of wastes, unregulated exploitation of marine resources and climate change. Appropriate compensating and mitigating mechanisms should be in place to obviate the risk of marine biodiversity erosion and the extinction of several species. Adopting the concept of Blue Economy, urbanisation with positive connotations, strict adherence to sustainable, low carbon and environment-friendly technologies to ocean development initiatives, are some means that can bring a world of change to the global way of life.

'Sagarmala' Initiative

India is a maritime nation, not just by historical traditions but also owing to its geophysical configuration and geopolitical circumstances.[3] It is the sixteenth largest maritime country in the world, and has a coastline of about 7,516 kilometres. With 13 maritime states and island territories, India probably has more seafaring people than the population of most European countries.[4] As a result, it is as dependent on the seas as any island nation for its economic growth and prosperity.[5] Indian ports and shipping industry play a vital role in sustaining the growth in trade and commerce and almost 95 per cent of India's trade is carried on by sea. India has 12 major and 200 notified minor and intermediate ports. Further, the cargo traffic which recorded 1,052 Million Metric Tonnes (MMT) in 2015, is expected to reach 1,758 MMT by 2017.[6]

Coastal shipping is of immense relevance for India, owing to the long peninsular coastline. In the recent years, rising delays and costs due to high road and rail congestion has been driving companies to consider coastal shipping to transport goods. The coastal shipping in India is significantly underutilised as compared to other emerging and developed countries. The share of coastal shipping in the overall cargo movement, i.e. 7 per cent, is extremely low as compared to that of the United States, some European and many Asian countries. The contribution of India's inland water transport cargo traffic to the logistics market is also significantly low at 0.5 per cent as compared to China at 8.7 per cent and Europe at 7 per cent.[7]

The 'Sagarmala' initiatives introduced by the Indian government include port-led industrialisation, urbanisation, coastal tourism and recreational activities, Short-sea shipping, coastal shipping, Inland Waterways

Transportation (IWT), shipbuilding, ship repair and ship recycling, logistics parks, warehousing, maritime zones/services, integration with hinterland hubs, offshore storage, drilling platforms, specialisation of ports in certain economic activities such as energy, containers, chemicals, coal, agro products, etc., offshore renewable energy projects with base ports for installations, modernising the existing ports and development of new ports. This strategy incorporates both port-led development viz. port-led direct development and port-led indirect development.[8] Project Sagarmala once implemented can potentially improve the trade and commerce across India and has the potential to revolutionise maritime logistics and port-led development in India.

Fisheries and Aquaculture

India has a long coastline, large areas under estuaries, backwaters, lagoons, etc., that are conducive for developing capture as well as culture fisheries. Fisheries and aquaculture constitutes one of the main sectors of food production. It provides occupation to about 14 million people engaged in different activities. According to the NFDB 2016 report, fish production in India constitutes about 6.3 per cent of the global fish production, whilst contributing to 1.1 per cent of the GDP and 5.15 per cent of the agricultural GDP.

Globally, India is ranked third in fisheries (behind China and Indonesia) and second in aquaculture behind China. Freshwater aquaculture continues to form a major share of the aquaculture production, with a contribution of over 95 per cent in terms of quantity. With an annual growth rate of over 6 per cent during the last decade, the sector has experienced high growth rates than other food-producing sectors. Though India has huge aquaculture resources, only 14 per cent of the potential water resources are being utilised, implying a huge potential for the growth of this sector.[9]

As regards deep sea fishing, India still lacks a proper blue water fishing fleet and the deep sea fishing industry is still at a nascent stage. The industry requires access to high technology fishing gadgets, gear and equipment which the native fishermen can ill-afford and hence, they have to rely on artisanal subsistence fishing. Additionally, owing to a lack of fish processing and cold storage facility at most of the fish-landing sites and fishing villages, the fishermen are forced to sell their yield immediately on landing ashore to the local agents, on a first-come-first-served basis and do not get the value for the yield.

Oil and Mineral Exploration

In 2015, India had an estimate oil reserve of 763.476 MMT and 1488.73 BCM of natural gas. The total number of exploratory and development wells and meterage drilled in offshore areas during 2014–15 was 1,352 thousand metres. Crude Oil production during 2014–15 was 37.461 MMT which is 0.87 per cent lower than 37.788 MMT produced during 2013–14. The refining capacity stood at 215.066 MMTPA in 2015 and by the end of 12th Five Year Plan, the refinery capacity is expected to reach 307.366 MMTPA. India exported 63.928 MMT of petroleum products against the imports of 20.423 MMT during 2014–15. Seabed resources of these areas and of the Exclusive Economic Zone (EEZ) covering about 1.87 million sq. kms, have also come to light in recent years[10]

A Geological Survey of India (GSI) study of the EEZ and territorial waters along the east and west coast of India has indicated the presence of heavy mineral placers such as limonite, ilmenite, gernet, monazite, zircon, rutile and magnetite[11] which are available in the form of polymetallic nodules. With their ever-increasing demand and diminishing land resources, these polymetallic nodules under the sea are being regarded as prospective alternatives. If the legal continental shelf is taken into account, the total offshore area would become about 3.09 million square kilometres.

India's mineral sector can be characterised as one of immense potential that is waiting to be tapped. Exploration to uncover this potential is crucial. An area of 1,50,000 square kilometres in the central Indian Ocean was allocated in 2002 by the International Seabed Authority (ISA) – the international body that controls and governs the exploration and extraction of minerals under the seabed. The closest estimates indicate 380 million tonnes of nodules in the retained Indian Pioneer area. However, following years of exploratory inaction and the lack of initiative, mining rights in many of the blocks in the south-west Indian Ocean had to be surrendered to the ISA. The Chinese dominance in the area of rare earth metals and sea-bed mining superiority is a cause of concern till India develops the capability to explore polymetallic nodules that contain rare earth metals.[12]

Leisure and Tourism

Tourism is one of the most exciting and progressive industries in the world. The willingness of the tourists for relaxation, enjoyment and leisure has increased and tourism has emerged as the largest global industry of the 20th

Century and is projected to grow even faster in the 21st Century.[13] India has immense possibilities of growth in the tourism sector with vast cultural and religious heritage and varied natural attractions. India is blessed with tourism destinations and products catering to the tastes and preferences of global tourists and travellers.

At the same time, the coastal areas are under very high population pressures due to the rapid urbanisation processes. More than half of the world's population lives in coastal areas (within 60 kilometres from the sea) and this number is on the rise (UNEP, 2014). Today, tourism is the largest service industry in India, with a contribution of 6.23 per cent to the national GDP and providing 8.78 per cent of the total employment. India receives more than 8 million foreign tourists annually along with 1,432 million domestic tourist visits.[14] The tourism industry generated nearly US$ 21.07 billion in 2015 and with the right policy measures and directives together with the Blue Economy initiatives, it is expected to increase to US$ 270 billion by 2026 at a 7.5 per cent annual growth rate.[15]

Challenges to the Blue Economy

To understand the constituents of the Blue Economy is a challenge in itself. Measuring the damage to the ocean environment and putting value to depleted biodiversity/fish-stocks of the oceans is a complex task. One of the characteristics of the sea is, that, it is cross-sectoral, often cross-jurisdictional, and trans-boundary.[16] It requires a real-time understanding of what happens at sea. Maritime domain awareness, information and technology sharing, remote sensing, ocean data acquisition devices, etc., are some of the key ocean surveillance mechanisms. They not only give a picture about the activities at sea, but also about the challenges that need to be tackled. Some of the challenges likely to affect a state implementing the Blue Economy are as follows:

Climate Change

Oceans and climate change are inextricably linked as the oceans play a role in mitigating climate change by serving as a major heat and carbon sink.[17] Marked by several similar characteristics, the elevation of the sea, the bleaching of corals, an increase in the temperature of the oceans among others, both climate change and environmental mismanagement have become growing threats to the integrity, condition, and sustainability of the aquatic and marine resources on which the Blue Economy is based. Although subject to local variations, many areas of the world, including India, have seen impacts of climate change that include increasing floods and droughts, erratic and extreme weather, distinct seasonal changes, sea level rise, coastal erosion, saltwater intrusion, warming sea waters, ocean acidification, coral bleaching, and an upsurge of invasive species as a result of global warming. These impacts worldwide are expected to worsen. They have already had negative effects on livelihoods which are dependent on ocean and freshwater resources, causing internal migrations, with potential for conflict, as well as putting at risk critical infrastructure and transportation systems.[18]

Marine Litter and Pollution

The oceans are littered with debris, threatening marine life and in some places concentrating in massive "garbage patches". Vast quantities of solid and chemical waste from human activities are continually dumped and reach into the sea, and plastics, sewage, oil and toxins accumulate in food webs. Pollution can poison marine life and decimate the marine environment. More than 250 different animal species - including seabirds, turtles, seals, sea lions, whales and fish – have been documented as having ingested marine debris or suffered from entanglement in the marine debris.[19] The run-off from land and harmful algae blooms have sapped the oxygen from water, creating marine dead zones around the world where fish and other marine life cannot thrive. Marine pollution is a global problem that threatens wildlife and presents health and safety concerns for humans as a polluted ocean will not be able to sustain marine biodiversity and thus is a challenge for the Blue Economy.

Illegal, Unreported and Unregulated Fishing and Overfishing

A significant percentage of the key fish stocks are overfished and/or depleted. Certain fishing practices also cause damage to the ocean itself, including excessive by-catch and destruction of vulnerable ecosystems and habitats. Illegal, Unreported, and Unregulated (IUU) fishing plagues the fisheries worldwide. The IUU fishing also exacerbates the problem of overfishing because the IUU vessels operate in marine protected areas where a total fishing ban has been imposed[20] (World Ocean Review, 2016). These vessels also pay little or no heed to fisheries management plans which are intended to conserve the overexploited or depleted stocks. The main reason why IUU fishing is a particularly critical issue today is that many fish stocks have already been overexploited by legalised fishing activities and the IUU fishing only puts such fish stocks under additional pressure. The IUU and overfishing are therefore a grave challenge to the Blue Economy.

Maritime Boundary Disputes

The overlapping jurisdictional claims and offshore boundary disputes are serious issues and States seek to secure the maximum maritime entitlements for themselves. The motivation behind delimitation efforts has unsurprisingly, generally been resource/demand for energy.[21] Traditionally, access to fishery resources represented the key issue; however, energy considerations now tend to be the primary concern. The 1982 UNLOS provides greater flexibility to States to demarcate maritime boundaries. States usually resort to a equidistance line or equitable solution, if there are no historic titles or special circumstances to resolve the boundary issues. However, no preferred method of demarcation has been indicated, thus the 'rules' are currently open to conflicting interpretation. Additionally, sovereignty disputes, especially over islands, have complicated matters and resulted in a number of conflicting claims to maritime space, and are extremely hard to resolve. Law enforcement is undoubtedly an expression of the sovereignty of the State within its territory and of its sovereign rights and jurisdiction within its contiguous zone, the EEZ and the continental shelf. Unresolved boundary disputes between maritime States cause insecurity to the people living or using the ocean/ maritime property, besides acting as a trigger for unlawful activities, which flourish against the backdrop of law enforcement agencies pitted against each other under the pressure of governance. The Blue Economy business in such waters will not be cost-effective due to insecurity.

Maritime Perils

The increasing globalisation of trade and commerce has also brought with it increased globalised threat dimensions taking forms hitherto unconceived of.[22] Terror attacks against maritime targets in the past have been quite diverse. Maritime shipping is vulnerable considering the growing volume of shipping traffic, low crew complement and restricted speeds due to congested choke points. Containerisation has also brought the risk of proliferation of Weapon of Mass Destruction (WMD), and biological or chemical attack. Offshore oil fields or oil platforms are also high interest targets as the destruction of these would put the world economy in chaos, notwithstanding the environmental catastrophe. Ports are extremely attractive targets for maritime terrorism. Attacks can be easily launched against a ship or port facility from land, sea, or even from underwater. Piracy and armed robbery also pose a major threat to sea navigation. All though piracy is on the decline in Somalia, however, the threats to shipping remain. Somali piracy operations were sophisticated as they were conducted by ex-militiamen who got assistance from technical experts and could operate hi-tech equipment such as GPS systems assisted with navigation and the detection of shipping targets. These men together with ex-fishermen had intimate knowledge of the sea, and formed a potent assemblage. Floating armouries as a result of piracy are also a cause of concern. These armouries, if not regulated and monitored, could be taken over by inimical forces and threaten maritime security and good order at sea. Maritime shipping and infrastructure, being one of the sources of a nation's commercial interest, qualifies as a potential and high value target of interest. No economy can progress nor attract investment in an insecure environment as insurance costs, facility security, maintenance costs, etc., may far outrun the cost advantage and thus the produce may not be able to compete in the market. Thus, maritime perils are detrimental to the growth of the Blue Economy.

Overcoming Challenges

The challenges discussed above are linked to maintaining a good order at sea. This requires clear and transparent regulations, effective round-the-clock monitoring, timely and strict enforcement of regulations, and above all, increased awareness. The path to national prosperity and high growth lies in secure trade routes and in an unhindered access to the sources of economic well-being, i.e. foreign trade, commerce and natural resources, which is to say that the security of the maritime domain is key to sustaining the economic development and growth.

Blue Economy Security: Stakeholders Role

Ocean affairs involves many stakeholders, including government agencies, spanning into science, environment, ecology, conservation, resources (living and non-living), fisheries, oil and gas, technology, transport, security and human resources. However, for the sustenance of the Blue Economy, a maritime domain protection against unlawful, illegal and antinational activities is mandatory. The maritime security prism is complex as it encompasses non-traditional security challenges such as natural disasters, illegal fishing, marine pollution, search and rescue, oil exploration, deep sea mining and climate change along with the traditional threats. With so many such threats in the seas, the management of maritime security or maritime security governance is integral to the Blue Economy.

Maritime Security is a combination of preventive and responsive measures, to protect the maritime domain against threats and intentional unlawful acts.[23] Ensuring maritime safety is also a sub-sect of maritime security operations. Safety and security go hand in hand and are integral to sound ocean governance and economic activity. There are many stakeholders who have a role to play in the sustenance of the Blue Economy; however, considering the maritime security governance as a whole, the following are the main stakeholders of Blue Economy security:

(a) DG Shipping

The Directorate General of Shipping of India is the national agency responsible for the implementation of the shipping policy and legislation so as to ensure the safety of life and ships at sea, prevention of marine pollution, promotion of maritime education and training in coordination with the International Maritime Organization, regulation of employment and welfare of seamen, development of coastal shipping, augmentation of shipping tonnage, examination and certification of Merchant Navy officers, supervision and control of the allied offices under its administrative jurisdiction, etc.[24] DG Shipping is the maritime administrator and conducts flag state inspection to ensure compliance with international conventions by the ships administered by it. It carries out strict Port State Control duties to ensure that foreign flagged vessels comply with the international conventions promulgated by the International Maritime Organization when they call at the Indian ports. The DG Shipping is also the national agency responsible for the implementation of the ISPS Code in India which is integral to the security of the Blue Economy.

(b) Indian Navy (IN)

The maritime realm is the legally used medium for power projection. The ability of a nation state to ensure free and full use of the seas, for trade, transportation and to meet resource needs is critical for a robust economic growth.[25] The maritime environment, accordingly, offers power and dominance to those who are strong at sea. This jostling ranges from wielding political influence, fishing and mineral mining rights, offshore economic activities like prospecting for oil and gas, etc., to denying potential adversaries the use of the seas by various 'hard' and 'soft' measures. A nation needs to develop adequate maritime power to effectively safeguard and progress its national interests. The navy is a versatile and potent instrument of national power, which can shape the maritime environment and safeguard national interests. The navy's military role is characterised by the threat or use of force at and from the sea. This includes the application of maritime power in both offensive operations against enemy forces, territory and trade, and defensive operations to protect our own forces, territory and trade.

(c) Indian Coast Guard (ICG)

The Indian Coast Guard is responsible for the surveillance of the Indian territorial waters and the Exclusive Economic Zone to prevent poaching, smuggling and other illegal activities. The ICG is the principal agency for protecting the ocean and offshore wealth, including oil, fish and minerals and also assists mariners in distress and safeguards life and property at sea and enforces maritime laws with respect to the sea, shipping, poaching, smuggling and narcotics. The Indian Coast Guard is thus responsible for executing/coordinating Search and Rescue (SAR) missions in the Indian Maritime Search and Rescue Region (ISRR). The Coast Guard is also the national agency for ensuring the marine environment security in India. It involves protection and preservation of the environment and the prevention and control of pollution. Being the national coordinator in oil spill response, it has a variety of responsibilities under the National Oil Spill Disaster Contingency Plan (NOS-DCP). It is the coordinator for oil spill response in the entire maritime zones of India with specific allocation for direct response functions in the maritime zones outside the port limits and notified areas around offshore oil facilities.

Among the three main stakeholders of maritime security, it is the Indian Coast Guard whose charter spans a number of sectors of the Blue Economy and can effectively support the national ocean development strategy particularly in the domains of marine environment, fisheries and scientific data collection.[26] The ICG is empowered and mandated through the Coast Guard Act, 1978 to support the national vision of developing the Blue Economy through environmental and ecological protection, prevent poaching and ensure protection of resources through enhanced maritime surveillance and patrolling the Maritime Zones of India.

ICG and the Blue Economy

The Indian Coast Guard is a dynamic organisation and its charter as a maritime law enforcement agency is to maintain constitutionalism in governance. Benign roles like SAR, anti-poaching, anti-smuggling, pollution response and as a constabulary "being on the beat" make it an excellent instrument of the state policy to protect, further and safeguard the Blue Economy within the maritime zones of India. The Blue Economy initiatives cannot be disassociated with the importance of continued inputs and analysis of oceans and to exploit the sea, it should be done via a sustainable manner to keep the good health of the oceans.

To successfully achieve these aims, an effective maritime awareness and information sharing arrangement must be in place to improve detection, collection and identification of capabilities that can be used to discover unknown or uncooperative maritime items of interest that represent threats, challenges and opportunities to the ocean users. India has made significant strides towards achieving the national Maritime Domain Awareness (MDA) after the Mumbai 26/11 attacks through the Indian Coast Guard by setting up a surveillance system called the Coastal Surveillance Network (CSN), a chain of static sensors having radars, Automatic Identification System (AIS), day/night cameras and met. sensors at 46 locations along the coastline and islands.[27] In order to achieve a near gap-free surveillance of the coastline, 38 additional radar stations and 08 mobile surveillance systems apart from VTMS connectivity are also envisaged as part of Phase II.

The Indian Coast Guard is networked under the Indian Navy's NC^{3}I network that links 51 Navy and Coast Guard stations, located along the coast as well as on island territories. The network provides these stations with coastal surveillance information obtained from various sensors such as the coastal radar chain of the Indian Coast Guard and automatic tracking systems as well as electro-optical cameras.[28] The MDA is required to enhance the protection of the maritime infrastructure by sharing and safeguarding the relevant information to protect the critical nodes and limiting access to those with validated permission and relevant roles and responsibilities; to maximise the legitimate use of the maritime domain by advanced maritime information sharing and an architecture to safeguard the infrastructure and processes to meet future validated requirements besides the stakeholders.[29] Integrating the space element, aerial surveillance, underwater and surface sensors into the ICG Maritime Domain Awareness architecture will ensure a multidimensional monitoring and collation of inputs towards overall ocean governance and good order at sea, which are important requisites for sustaining the Blue Economy.

Conclusion

The Blue Economy envisages a sustainable and inclusive approach, since it promotes growth and employment. It fosters innovation, both by supporting the development of new industries and by encouraging new ideas in established sectors such as fishing. It allows economies to secure more value from their maritime zones, and opens up new opportunities for developing economies. An integrated approach applied to the Blue Economy is a sustainable answer to many of the challenges forecasted by the current megatrends such as population growth, resource scarcity, climate change, shifts in economic power and technological breakthroughs. However, it is important to keep in mind that the ocean resources cannot be exploited. Hence, a robust and effective ocean order and governance through effective maritime security can ensure confidence building measures for enabling the ocean industries to invest and carry out ocean-related activities without any insecurity.

The IUU and excessive fishing can be effectively countered through stringent Monitoring Control and Surveillance (MCS) regime of the fisheries. A strict enforcement of Code of Conduct of Responsible Fisheries (CCRF), advocated by the FAO, needs to be implemented. A vigorous MDA and MCS regime will ensure effective deterrence against unsustainable practices which ultimately may ensure the bluing of the oceans and the greening of the earth to mitigate the effects of climate change. Ocean surveillance is thus integral not only for ensuring effective maritime security deterrence but also for ensuring an enforcement of regulations/instruments both national and international and a close monitoring of the environmental and ecological health of the oceans. It is clearly evident that maritime security and the Blue Economy are intertwined and integral to each other. They are undoubtedly both sides of the same coin and neglecting any one of them will be at the risk of the other.

REFERENCES AND NOTES

1. The Economist Intelligence Unit, 'The Blue Economy: Growth, Opportunity and a Sustainable Ocean Economy', 2015. Retrieved from https://www.eiu perspectives .economist . com /sites /default /files/images /Blue percent20Economy _briefing percent 20 paper WOS2015.pdf

2. World Bank Environment Brief, 'Ocean, Fisheries and Coastal Economics', 2016. Retrieved from http://www.worldbank.org/en/topic/environment/brief/oceans
3. Singh, Sidharth, 'Security Outlook of Indian Ocean India's Geostrategic Interests in the IOR', *India Foundation Journal*, IV (4), 2016. Retrieved from http://www India foundation .in/wp-content/uploads/2016/10/India-Foundation-JOURNAL-July-Aug-2016.pdf
4. Dahiya, Omprakash. 'Maritime Security of India: Challenges and Opportunities', *Employment News*, Vol. XL (1), 2015. Retrieved from www.employmentnews.gov.in
5. Prakash, Arun, 'Maritime Security of India: Future Challenges', IDSA, 2013. Retrieved from http://www.idsa.in/keyspeeches/MaritimeSecurityOfIndiaFutureChallenges
6. India Brand Equity Foundation, 2016. Retrieved from http://www.ibef.org/industry/ports-india-shipping.aspx
7. AIVP, 'Insights into Indian Maritime Industry and Ports in India', 2016. Retrieved from http://www.aivp.org/wp-content/uploads/2016/06/suresh _trishala _indian _ maritime industry.pdf
8. Press Information Bureau, 'Sagarmala: Concept and Implementation towards Blue Revolution', Press Release, 2015. Retrieved from http://pib.nic.in/newsite /Print Release.aspx ?relid=117691
9. MPEDA, 'Indian Seafood Industry - An Overview', 2016. Retrieved from http:// www . indianseafoodexpo. com/indian_seafood_industries.php?a=2
10. 'Indian Petroleum and Natural Gas Statistics', Ministry of Petroleum & Natural Gas, GoI, 2015. Retrieved from http://www.indiaenvironmentportal.org. in/files /file /png stat percent202014-15.pdf
11. Indian Bureau of Mines, Min. of Mines, GoI, 'Ebook on Minerals', 2016. Retrieved from http://pib.nic.in/eventsite/ebook/pdf/Ebookonmineralsector.pdf
12. Singh, Abhijit, 'India's Deep Sea Mining gets a Fillip', Institute for Defence Studies and Analysis, 2013. Retrieved fromhttp://www.idsa.in/idsacomments/Indias deep sea mining capability getsafillip_asingh_011113
13. Sharma, A., 'Economical Growth of Tourism in India', *IJECR*, vol. 3(1), 2013, pp.115-118.
14. 'India Tourism Statistics', Ministry of Tourism, Govt of India, 2015. Retrieved from http:/ / tourism.gov.in/market-research-and-statistics
15. World Travel and Tourism Council, 'Travel and Tourism: Economic Impact India', 2016. Retrieved from http://www.wttc.org/-/media/files/reports/economic-impact-research / countries -2016/india2016.pdf
16. Bueger, Christian, 'Maritime Security and the Blue Economy: Understanding the Link', 2016. Retrieved from http://www.westafricasecurity network.org/maritime-security-and-the-blue-economy-understanding-the-link/
17. Ocean Foundation, 'Ocean and Climate Change', 2016. Retrieved from https:// www. Oceanfdn.org/resources/ocean-and-climate-change
18. UNECA, 'UNECA Africa's Blue Economy: Policy Handbook', 2016. Retrieved from http:/ /www.uneca.org/sites/default/files/PublicationFiles/blue-eco-policy-hand book _eng _1 nov.pdf
19. Jones, Kerri-Ann, 'Ocean Sustainability: U.S. Perspectives', The Economics of the Oceans Conference, 2013. Retrieved from https://www.state.gov/e/oes/rls/remarks /2013 / 218689.htm
20. World Ocean Review, 'Fisheries', 2016. Retrieved from http://world ocean review .com/ en /wor-2/fisheries/illegal-fishing/
21. Schofield, C. H., 'Cooperative Mechanisms and Maritime Security in Areas of Overlapping

Claims to Maritime Jurisdiction' in *Capacity Building for Maritime Security Cooperation in the Asia-Pacific: A Selection of Papers presented at the CSCAP Study Group Meeting on Maritime Security Cooperation, December 2004, Kunming, China and April 2005, Delhi, India*P. Cozens and J. Mossop (eds), Wellington, New Zealand: Centre for Strategic Studies, 2005.

22. Kumar, D., 'Asymmetric Threats and Their Challenges to Freedom of Navigation',Malmo Study, Sweden: World Maritime University, 2010.

23. The Wise Pen Final Report on Maritime Surveillance in Support of the CSDP gives the definition of Maritime Security. Accessedhttp://www.realinstitutoelcano.org/wps/wcm/connect/44d0718042982ce1bb66bb24ab1546e8/The_wise_pen_team_final_report_april_2010.pdf?MOD=AJPERES&CACHEID=44d0718042982ce1bb66bb24ab1546e8

24. Details about DG Shipping obtained from website http://www.dgshipping.gov.in. Accessed 15 Dec 2016

25. 'Indian Maritime Doctrine', Indian Navy, Min of Defence, GoI, 2009. Retrieved from https://www.indiannavy.nic.in/sites/default/files/Indian-Maritime-Doctrine-2009-Updated-12Feb16.pdf

26. Sakhuja, Vijay, 'Blue Economy and Indian Coast Guard', *South Asia Defence and Strategic Review*, Vol. 10, 2016, p.24. Retrieved from http://www.defstrat.com/current-issues/blue-economy-and-indian-coast-guard

27. Retrieved from http://www.indiancoastguard.gov.in/content/1727_3_CoastalSecurity.aspx on 10 Dec 16

28. Press Information Bureau, Government of India, Ministry of Defence, 'Coastal Security Network Must Ensure Zero Tolerance to Error: Parrikar', http://pib.nic.in/newsite/PrintRelease.aspx?relid=111697

29. 'U.S. National Strategy for Maritime Security: National Plan to Achieve Maritime Domain Awareness' 2005. Retrieved July 11, 2010 from the World Wide Web: dhs.gov/sites/default/files/publications/HSPD_MDAPlan_0.pdf

10

Streamlining Maritime Administration for Providing Sustainable Services

Pradeep Chauhan

This chapter presents maritime administration as a term that encompasses not just mercantile shipping, but the entire gamut of activities undertaken in maritime spaces. This chapter avers that India's national-level maritime administration must reflect an evident outflow from the apex international body that deals with this subject, viz., the UN 'High-Level Political Forum (HLPF) on Sustainable Development'. Also, it addresses India's committed targets on Goal 14 – including those to be achieved by 2025 and 2030. The chapter goes on to dwell upon a few of the more significant maritime activities – mercantile shipping, maritime administration vis-à-vis fisheries, and maritime administration vis-à-vis security. It outlines four international models of maritime administration organisations and task – in respect of Australia, Canada, the European Union and the USA, on the basis of which the chapter specifies the Twenty-One Tasks that the streamlined Indian maritime administration must undertake as a minimum. The chapter thereafter dilates at length upon the organisational structure, legislative support and coordination functions required, in both textual and schematic formats, in order to reinvent the Indian Maritime Administration. The chapter concludes by emphasising the absolute criticality of proactively creating and promoting maritime awareness, maritime education and maritime public relations.

Introduction

In India, the expression 'Maritime Administration' evokes a nearly universal mental image of a civilian organisational structure that deals with the issues and concerns of mercantile shipping, i.e., seaborne trade. This stubbornly embedded association of maritime administration with mercantile shipping tends to hijack, if not derail, any attempt to widen the discussion to incorporate not just trade and shipping, but the entire gamut of activities undertaken by the users of those maritime spaces over which the Republic of India has varying degrees of sovereign jurisdiction, as also those maritime spaces that lie beyond the legal jurisdiction of any nation-state.

Stakeholders

There are numerous stakeholders whose individual and collective activities are sought to be administered such that they remain in consonance with India's pursuit of a Blue Economy. This noble pursuit conceptualises the oceans as 'development spaces' and seeks to integrate environmental conservation, sustainable use, oil and mineral wealth extraction, bio-prospecting, sustainable energy production and marine transport. It is important to note that the decoupling of socio-economic development from environmental degradation lies at the core of the Blue Economy. The concept recognises and places renewed emphasis on the critical need for national organisations as well as the international community to address effectively the sound management of resources in and beneath international waters by the further development and refinement of a national legislation that conforms to the international law and ocean governance mechanisms.

On the one hand, New Delhi is clear that India – a manifestly maritime State – is going to take its responsibilities in this regard with the utmost seriousness and sincerity. On the other, it wishes to not merely join, but to be in the vanguard of international efforts to ensure that each country takes its share of the responsibility to protect the 'high seas', which cover 64 per cent of the surface of our oceans and constitute more than 90 per cent of their volume. Given the sheer abundance of organisations and individuals that are involved in one or another manner with matters maritime, no one should be under any illusion that this is a simple matter. Even a cursory survey will identify the following stakeholders:

(i) The fishing community, comprising traditional, coastal, and deep-sea fishermen, as also multinational fishing conglomerates and fleets.

(ii) The mercantile shipping community, comprising a large variety of coastal shipping – from traditional sailing boats and country craft to large dhows propelled both by motor and sail, and motorised ships engaged in coastal trade – and larger ships, both Indian and foreign-flag, engaged in external merchandise trade.

(iii) The ancillary maritime community, incorporating a bewildering array of specialised craft of varying degrees of sophistication, such as dumb and self-propelled barges, diving vessels, ferries, pilot vessels, vessels engaged in buoyage, tugs, dredgers, salvage vessels, offshore exploration vessels, drilling vessels/ships, oil production and support vessels, etc.

(iv) The scientific community comprising, inter alia, oceanographers, ocean thermal and ocean mechanical engineers and technicians, hydrographers, ocean-hydrologists, marine biologists, earth scientists, mining companies, alternative energy seekers, space and remote-sensing scientists, and so on and so forth.

(v) The maritime leisure/tourism community, comprising once again, a staggering variety of craft propelled by oars, sails, motors, engines and all combinations thereof.

(vi) The maritime safety and security community, incorporating departments and organisations related to the Indian Coast Guard, the Chief Hydrographer to the Government of India, state and central paramilitary and police forces, customs and revenue, immigration, health and sanitation, pollution detection and control, Directorate General of Lighthouses and Lightships, etc.

(vii) Ships, vessels, surface-, sub-surface, air- and hybrid-craft and personnel of the Indian Navy, and, to a lesser extent, the Indian Air Force and the Indian Army.

The organisational structures that partially or wholly (and with varying degrees of efficiency) administer the activities of these various maritime communities are in no small number either. They include, amongst others, the following:

(a) *International Organisations:* The United Nations Commission on Sustainable Development, the International Maritime Organisation (IMO), the International Court of Justice (ICJ), the International Tribunal for the Law of the Sea (ITLOS), the International Seabed Authority (ISA), the Antarctic Treaty Organisation, the International

Hydrographic Organisation (IHO), regional hydrographic organisations. etc.

(b) *Multinational Organisations:* The International Maritime Bureau (IMB), Marine Insurance/Assurance Companies and Associations, Classification Societies, etc.

(c) *National (Central Government) Organisations:*

- The Ministries of:
  - Defence (MoD)
  - Shipping (MoS)
  - External Affairs (MEA)
  - Home Affairs (MHA)
  - Finance (MoF)
  - Health & Family Welfare
  - Agriculture & Farmers Welfare (MoAFW)
  - Petroleum and Natural Gas (MoPNG)
  - New and Renewable Energy (MNRE)
  - Earth Sciences (MoES)
  - Law (MoL)
  - Environment, Forest and Climate Change (MoEFCC)
- Department of Atomic Energy
- Department of Space
- Office of the Chief Hydrographer to the Government of India
- The Indian Navy
- The Indian Air Force
- The Indian Coast Guard
- State Marine Police
- Central Industrial Security Force
- Bureau of Immigration
- Department of Animal Husbandry Dairying & Fisheries (DADF)
- Director General (Shipping)
- 13 × Major Port Trusts
- 200 × Notified Minor Ports:
  - Gujarat Maritime Shipping Board (for Minor Ports in the states of, and)
  - Secretary (Ports), UT of Daman and Diu
  - Maharashtra Maritime Shipping Board
  - Captain of Ports Department, Government of Goa
  - Department of Public Works, Ports and Inland Water Transport, Government of Karnataka

- Director of Ports, Government of Kerala
- Director of Port, Government of Puducherry
- Secretary (Transport), UT of Lakshadweep
- Tamil Nadu Maritime Shipping Board
- Director of Ports, Government of Andhra Pradesh
- Directorate of Ports & Inland Water Transport, Government of Odisha

- □ Port Health Officers
- □ Directorate General of Lighthouses and Lightships
- □ Directorate General of Hydrocarbons
- □ Gas Authority of India Limited (GAIL)
- □ Department of Ocean Development (DOD), Goa
- □ National Institute of Oceanography (NIO), Goa
- □ Department of Ocean Technology, Tamil Nadu
- □ National Centre for Antarctic and Ocean Research (NCAOR), Goa
- □ Indian National Centre for Ocean Information Services (INCOIS), Hyderabad

Each of the foregoing organisations, ministries, departments, services, trusts and agencies play a role in the business of the management, exploitation, conservation, legislation and security of the maritime spaces of interest. It would be immediately apparent that any plan for the single-point administration of the activities of all these disparate structures is critical. If the aims of the Blue Economy are to be pursued and attained in any coherent (and hence predictable and controllable) manner, the need for some form of efficient and effective coordination (administration) is obvious.

International Formal Commitments

It would be prudent for a national-level maritime administration to reflect an evident outflow from the apex international body that deals with this complex subject, namely the United Nations 'High-Level Political Forum (HLPF) on Sustainable Development', which has, since 2013, replaced the '1992 Commission on Sustainable Development' (CSD). The creation of the HLPF as a successor to the CSD had been agreed to by the Member States at the United Nations Conference on Sustainable Development (Rio+20) in 2012, so as to ensure that sustainable development tops the agenda of the highest levels of government and is embraced by all actors.[1] The globally-agreed goals for sustainable development for the post-2015 development agenda (covering the period 2015 to 2030) have been set forth in the UN document entitled, "Transforming our World: the 2030 Agenda for Sustainable Development".[2] This document, known commonly as the '2030 Agenda', was formally agreed to by the UN General Assembly from September 25 to 27 2015 and articulates 15 specific goals and 169 targets, which have now replaced the Millennium Development Goals. Of these 15 goals, it is Goal 14: "*Conserve and sustainably use the oceans, seas and marine resources for sustainable development*".

An important aspect of Agenda 14 is that as part of the follow-up and review mechanisms of the '2030 Agenda for Sustainable Development', developed and developing member states are required to conduct regular and inclusive reviews of progress at the national and subnational levels. These national reviews are to be country-led and country-driven and are to provide a platform for partnerships, including through the participation of major groups and other relevant stakeholders. These national reviews are specifically expected to serve as a basis for regular reviews by the HLPF, which began in 2016. This review will be undertaken during the next meeting of the HLPF, under the auspices of the UN Economic and Social Council (ECOSOC), from July 10 to 19, 2017. The 2017 meeting will incorporate a three-day ministerial meeting (scheduled every four years) that will be attended by Heads of State/Government.[3] It will undertake an in-depth review of seven of the 15 Sustainable Development Goals that have been agreed upon, amongst the seven is Goal 14. In preparation for this review, a special UN Conference, "The Ocean Conference" is being held from June 05 to 09, 2017, in New York.[4]

Mercantile Shipping

The opening arguments of this chapter, cautioning the reader against a grossly narrow approach in considering that maritime administration is concerned solely with mercantile shipping, i.e., maritime merchandise trade, are reiterated. However, it is also beyond a shadow of doubt that the administration of mercantile shipping is a vital piece of the maritime administration jigsaw puzzle. International shipping transports more than 80 per cent of the global trade to people and communities all over the world, for shipping remains the most efficient and cost-effective method of international transportation of most goods.

International Maritime Organization (IMO)

Insofar as the mercantile shipping goes, the maritime administration at the international and national levels is linked through the International Maritime Organization (IMO), "which is the United Nations specialized agency with responsibility for the safety and security of shipping and the prevention of marine pollution by ships. Its main role is to create a regulatory framework for the shipping industry that is fair and effective, universally adopted and universally implemented".[5] So as to ensure that this vital sector grows and yet remains safe, environmentally sound, energy efficient and secure, the IMO regulatory measures cover all aspects of international shipping, including ship design, construction, equipment, manning, operation and disposal. Additionally, the IMO ensures that ship operators cannot address their financial issues by simply cutting corners and compromising on safety, security and environmental performance. In so doing, the IMO seeks to make the sustenance and growth of mercantile shipping a feature of human endeavour that is entirely in consonance with the aims of the Blue Economy concept.

International Maritime Bureau (IMB)

It is necessary to distinguish between the IMO and the IMB – the International Maritime Bureau. The latter is a specialised agency, not of the UN but of the International Chamber of Commerce (ICC). As such, it is a multinational organisation established (in 1981) by various mercantile shipping stakeholders, with the aim of protecting the integrity of international trade by seeking out fraud and malpractice. It positions itself as a focal point in the fight against all types of maritime crime. It is reiterated that while the IMB does enjoy a formal encouragement from the IMO and has an MOU with the World Customs Organization (WCO) and observer status with Interpol (ICPO), it is not an intergovernmental body as is the IMO, but a non-governmental one[6] that seeks to protect the narrower, commercial interests of mercantile shipping.

Maritime Administration vis-à-vis Indian Mercantile Shipping

As far as India is concerned, the country had a closed economy until the beginning of the 1990s. It has now seen over a quarter of a century of acceleration in its opening-up and integration with the global economy, characterised by a sharp relaxation of government controls, with a substantial role for the private sector as the engine of growth in a free market framework, nurturing competitive forces in order to boost efficiency. The increasing dependence upon international trade – over 30 per cent of which is accounted for by sea freight[7] – is a ubiquitous feature of globalisation and India is no exception to this trend. Table 1, which is drawn from World Bank data (in current U.S. dollars), clearly shows that India's (and China's) merchandise trade, as a percentage of the GDP (also known as the 'Openness Index') has been rising sharply since the 1980s. India's Trade-to-GDP ratio has jumped from 14 per cent in 1991 to the present average decadal value of 37 per cent – significantly higher than the global average of about 24 per cent.

Table 1. Trade-to-GDP Ratio (Openness Index)

Year	India	China	Pakistan	UK	USA
2015	31.5	35.9	24.6	38.1	21.2
2014	38.5	41.6	29.6	40.0	23.2
2013	41.9	43.8	30.2	44.3	23.5
2012	43.1	45.7	30.6	44.4	24.0
2011	42.3	48.6	32.5	45.6	24.2
2010	33.7	49.2	33.4	41.9	21.7
2009	30.9	43.6	29.3	37.8	18.5
2008	42.1	56.2	36.8	40.5	23.5
2007	30.6	61.8	33.1	36.4	21.9
2006	31.6	64.5	34.1	41.1	21.2
2005	29.1	62.7	37.8	37.6	20.1
2004	24.4	59.5	32.0	35.6	19.1
2003	21.3	51.6	30.0	36.3	17.6
2002	20.2	42.5	29.2	38.3	17.2
2001	19.0	38.3	26.9	40.1	18.0
2000	19.7	39.4	26.9	40.7	19.8
1990	12.7	32.2	32.6	38.3	15.2
1980	12.4	20.1	33.6	39.8	16.9

The world economic environment has a direct impact upon this Trade-to-GDP Ratio. In periods of economic slowdowns, exports as well as imports tend to decline and this adversely impacts a country's Openness Index. This is the situation that has prevailed through the latter half of the Fiscal Year (FY) 2015–16, wherein "growth in emerging market and developing economies declined for the fifth consecutive year. As a result, overall global economic activity remained subdued in 2015".[8] On the other hand, as the world economy regains momentum, the Index will be positively impacted. For the present, the outlook is only marginally encouraging and the current IMF projections are that growth in the global economy will improve from 3.1 per cent in 2015 to 3.4 per cent in 2017. That said, and the instantaneous variations in the specific values of the Trade-to-GDP Ratio notwithstanding, it is nevertheless clear that over the past few decades, as the economy of India (and China) have become increasingly integrated with that of the world, the external dimension of each economy is now substantial. India's present GDP (nominal) stands at some at US$ 2.25 trillion.[9] Even in 2015, which was a very lean year for world trade, India's merchandise exports amounted to US$ 271.745 billion, while its imports were US$ 391.176 billion, yielding a total merchandise trade of US$ 662.921 billion (approximately US$ 0.663 trillion).[10] This represents just under a third of the country's GDP. 95 per cent of this[11] (about US$ 630 billion) is seaborne. Thus, if one were to simplistically average this value over 12 months, it may be seen that, in any given month of 2015, some US$ 52.5 billion worth of India's merchandise goods (constituting imports as well as exports) were to be found upon the seas.

The optimal administration of merchantmen calling at the Indian ports is, therefore, a matter of considerable importance. The fact that India's merchant fleet figures in the global 'top-twenty' is a matter of some satisfaction, but ought not to generate undue euphoria – not the least because the fleet's size and capacity is a sum of everything from tugboats to super-tankers. The largest fleets by 'flag of registration' in 2015 were those of Panama, Liberia, the Marshall Islands, Hong Kong (China) and Singapore. Together, these top five registries account for 56.5 per cent of the world tonnage (the fact that the top three are actually 'Flags of Convenience' is a separate problem with significant connotations for maritime security). Of the 1,204 ships in the Indian Register on December 31, 2014, only some 30 per cent, i.e., 358 vessels were being deployed for overseas trade. The vast bulk – just over 70 per cent – was engaged in coastal trade. While that might by itself indicate a thriving coastal seaborne trade, this is far from the case, because the tonnage carried by this 70 per cent is a mere 11.8 per cent of the total Gross Tonnage of the Indian Fleet. In other words, only 30 per cent of Indian shipbottoms account for 88.2 per cent of the Gross Tonnage! Barely 8 per cent of India's own merchandise trade is carried on Indian ships, reducing the effect of administrative measures that India may take.

Another factor that impacts maritime administration is the age of the merchant fleet. The average age of the Indian shipping fleet as on December 31, 2014 is shown in Table 2.[12]

Table 2. Average Age of the Indian Shipping Fleet

Category	0-5 Years		6-10 Years		11-15 years		16-20 years		> 20 years		Total	
	Total	%	Total	%	Total	%	Total	%	Total	%	Total	%
Coastal	177	21.0	110	13.0	100	11.8	94	11.1	365	43.1	846	100
Overseas	104	29.1	64	17.9	37	10.3	39	10.9	115	31.8	358	100
Total	281	23.3	174	14.5	137	11.4	133	11.0	479	39.8	1204	100

It is clear from the foregoing tabulation that in both categories of maritime trade (coastal and overseas), the Indian-flag shipping has a significant proportion (54.2 per cent and 50.8 per cent respectively) of vessels that are more than 15 years old and whose construction and outfitting technology is out of step with modern global safety and efficiency practices. A merchant-fleet comprising largely of old ships is likely to be more violative of environmental and safety standards laid down by the IMO. These are largely enforced via 'Port-State Control' (PSC), whereby a port-state carries out inspections of foreign-flag shipping to ensure compliance with extant regulations promulgated by the IMO. Specific authority is provided through various 'Memorandums of Understanding' (MOUs), the first of which was the Paris MOU of 1982. Since then, a number of regional MOUs have been signed. The more well known of these are the Tokyo MOU of 1993, the Acuerdo Latino or Acuerdo de Viña del Mar (South and Central America) of 1992, the Caribbean MOU of 1996, the Mediterranean MOU of 1997, the Abuja MOU of 1999 (West and Central Atlantic Africa), the Indian Ocean MOU of 1999, the Black Sea MOU of 2000 and the Riyadh MOU (Persian Gulf) of 2005.

The United States Coast Guard is an independent PSC authority for the USA. The Indian-flag ships have, from time to time, been detained by various PSC authorities, causing financial loss and time-delays to the shipping companies concerned. For instance, between 2012 and 2014, 16 Indian-flag ships had been detained under the Tokyo MOU and while Indian-flag shipping is not in the 'Black List', it certainly does figure in the 'Grey List'[13] (The Asia-Pacific PSC (Tokyo MOU)'s 'Black', 'Grey' and 'White' list for flag State performance is established annually taking account of the inspection and detention history of the preceding three calendar years and is adopted by the Tokyo MOU Committee to publish in the Annual Report). The same is true of the Paris MOU. The Indian merchant ships have been detained under the Indian Ocean MOU as well. In several cases, these detentions and the fiscal and time penalties that accrue therefrom are functions of the advanced age of the ship and the consequent lack of technological advancements on board.

Maritime Administration vis-à-vis Fishing

The 2014 "Report of the Expert Committee Constituted for Comprehensive Review of the Deep Sea Fishing Policy and Guidelines" pegs the current resource-potential of the Indian EEZ, in terms of fish, at 4.41 million metric tonnes.[14] The report, which is categoric in its commitment to the Blue Economy, emphasises that production from the near-shore waters in the Indian Exclusive Economic Zone (EEZ) has now plateaued and there is minimal scope for further hikes in production. It presses for a maritime administration that would ensure sustainable exploitation of resources in our near-shore waters so as to ensure that commercial fisheries do not collapse altogether. It strongly urges that the thrust of fishing endeavours be shifted from coastal waters to the deep sea. It also seeks a major revamping of the extant guidelines in respect of deep-sea fishing activities, so that resources, especially tuna and tuna-like species, may be optimally harvested from the Indian EEZ.

In terms of monetary value, there is a clear need for the streamlining of the maritime administration related to the country's fisheries industry. Data released by the Marine Products Export Development Authority (MPEDA) indicates that India's seafood product exports were about 30,421 crore rupees (US$ 4,688 million) during the financial year 2015–16, with Southeast Asia constituting the principal destination, accounting for 39 per cent of the total exports.[15]

The fishery resource potential of just one type of living resource within the Indian EEZ namely, oceanic tuna, is 2,13,000 tonnes with an estimated composition of yellowfin at 54 per cent, skipjack at 40 per cent and big-eye at 6 per cent. The monetary value is valued at approximately INR 3,000 crores or US$ 500 million.[16] Yet, due to the lack of national enterprise and human expertise, there is an absence of specialised fishing fleets in India for the harvesting of tuna. Consequently, the landings of oceanic tuna in India have been negligible.[17] Indeed, as the 2014 Expert Committee put it quite unequivocally, "poor entrepreneurship in deep sea fishing, lack of endurance, use of old and outdated technologies, absence of R&D inputs in modernization of the fishing fleet are some of the hurdles yet to be crossed by the marine fisheries sector to make India a true deep sea fishing nation".[18]

In the absence of either Indian fishing fleets or adequate policing of its

EEZ by the Indian Coast Guard, the migratory stocks of tuna and tuna-like species are being caught by the fishing fleets of neighbouring tuna-fishing nations such as Maldives, Sri Lanka, Thailand and Indonesia. This is obviously a net loss of revenue to India. On the other hand, because these foreign fishing vessels/fleets are unconcerned about the conservation of the stocks or staying within the maximum sustainable yield of the fish, they tend to overfish the tuna. Tuna is a migratory species, if it senses that its survival as a species is seriously threatened, the tuna migratory path might well change to avoid large segments of the Indian EEZ. If this were to happen, even when – in the eventual future – Indian fishermen do go out to catch tuna within the Indian EEZ, there may be no tuna to catch! Therefore, any prolongation of the current leisurely and 'vaguely academic' approach to this situation is quite unaffordable. In other words, whether or not this resource is harvested, it must nevertheless be policed to ensure that it is not 'poached'. The Indian Coast Guard (ICG) is alive to its responsibility for effective policing in the country's EEZ. Yet, unfortunately, such is the pressure upon this young organisation from the demands of coastal security vis-á-vis the illegal immigration of arms and State-sponsored non-State actors that it is well-nigh impossible for the Coast Guard to concentrate adequate resources on preventive-policing of the EEZ.

The case for requisite assurance of safety, too, demands the same Coast Guard resources. Indian fishermen expect that the State will protect them while they are in the pursuit of their legitimate livelihood, namely, fishing. As and when deep sea fishing becomes an intensive Indian activity, assuring and affording Search and Rescue (SAR) cover to these fishermen in distant waters will become an essential feature of the Indian Coast Guard.

Because of the need to de-conflict 'policing' activities by the State agencies (as also safe passage for warships), from the needs of fishermen to legitimately earn their livelihood, it is necessary to take forward a key recommendation of the 2014 Expert Committee – and this is to do with 'Fish Aggregating Devices' (FAD). Since fishermen will go where the fish are, conflicts between the fishing community and security agencies have been a marked feature of the Indian discourse. On the one hand, the fishermen cite a loss of livelihood due to Indian naval and Coast Guard vessels cutting their nets. On the other, the security agencies accuse the fishermen of being unmindful of zoning or traffic-separation mechanisms, especially in waters in the vicinity of ports and harbours, and close to the coast, thereby causing a fouling of propellers, damage to sonar domes and other underwater sensors, obstruction to naval and coast guard units conducting essential exercises, and creating navigational hazards – especially to submarines. The FAD holds out great promise in a mutually beneficial resolution of this conflict. It is well known that epipelagic fish are fascinated with floating objects and aggregate in considerable numbers around such objects. The objects appear to provide a "visual stimulus in an optical void". This is true, not only for 'neritic' or 'coastal' fish (which is where the problem is currently being felt most acutely) but also of the 'oceanic' epipalegic fish – such as tuna – as well, which are found in the deeper reaches of the Indian EEZ (which is where the problem will increasingly be felt in the years to come). In fact, the Indian Ocean Island nation-states such as Mauritius and Seychelles have been using FAD for decades, as a means of ensuring a plentiful catch for each of their territorially sensitive fishing communities – with uniformly excellent results. The Department of Fisheries lays out the FAD for each village-group and promulgates the location accordingly. Each fishing group proceeds to its promulgated FAD and fishes away merrily, capitalising upon the innate curiosity of the epipalegic fish-species that congregate around the FAD. In this manner, each fishing group is segregated from the other in time and space, minimising conflicts at sea. Nor is this a case of merely folk-wisdom – it has been scientifically studied and meticulously recorded in a number of aquaculture journals. For instance, in one study of the efficacy of the FAD to attract bigeye tuna, yellowfin tuna and albacore tuna:

> "87 in number, two-hour acoustic surveys (radius 0.8 nautical mile, vertical range 0–500 m) around 17 fish aggregating devices (FADs) were conducted in French Polynesia between December 1995 and February 1997. Associated tuna densities were calculated using two different techniques: echo counting when the fish had sufficient distances from each other and echo integration when the fish swam close together (in schools)...... [The] surveys showed three different classes of aggregations: (1) 'deep scattered fish', observed 45 times, (2) 'intermediate scattered fish', observed 16 times, and (3) 'shallow schooling fish', observed 16 times. Sometimes aggregations of different classes were observed beneath the same FAD...."

It may reasonably be concluded that if the 'affected' parties – the Indian Navy, the Indian Coast Guard, the Department of Fisheries and the local fishing communities – were to lay out FAD in accordance with the principles of operational planning (something that the Indian Navy is exceptionally good at), the fish – and hence the fishermen – could be aggregated in areas away from the shipping lanes and away from naval exercise areas. It is simply astonishing that despite this subject having been broached on numerous occasions and in numerous fora with all concerned stakeholders present, its application continues to evade us.

While very substantial progress has been made in terms of regulation of fishing vessels, self-congratulation must be tempered by the recognition of the fact that the process has been commenced practically from scratch and that while progress has been substantial, it is by no means adequate. The 2014 Experts Committee noted the fact that within the Indian EEZ, "the domestic fleet.... operates in a legal vacuum, with no license or regulatory mechanism. This situation has arisen because the proposed Bill on Marine Fishing (Regulation and Management) of Fisheries in the EEZ is yet to see the light of the day".

Maritime Administration vis-à-vis Security

The 2030 Agenda specifically recognises the linkages between sustainable development and security, stating unequivocally, "sustainable development cannot be realized without peace and security; and peace and security will be at risk without sustainable development".[19] Even within the maritime context, it is not merely military security to which this averment refers, but to holistic maritime security (incorporating traditional/military as well as non-traditional/ human security). Holistic maritime security has been accurately described as "freedom from threats arising in or from the sea".[20] India's approach has its historical moorings in the famous "Common Security" report that had been authored as long ago as 1980, by the "Independent Commission on Disarmament and Security Issues" chaired by the late Prime Minister of Sweden, Mr Olaf Palme. This report emphatically drew attention to the alternative ways of thinking about peace and security by formally acknowledging that common security requires that people live in dignity and peace, that they have enough to eat, and are able to find work and live in a world without poverty. While military maritime security does, of course, continue to enjoy primacy for India, existing as it does in a world-system defined by Westphalian concepts of national sovereignty, maritime security is now firmly established within a new construct that incorporates military, political, economic, societal and environmental dimensions, and recognises the many linkages between them. Thus, threats to human security, such as religious extremism, international terrorism, drug and arms smuggling, demographic shifts – whether caused by migration or by other factors –human trafficking, environmental degradation, energy, food and water shortages, all now figure prominently as threats that are inseparable from the military ones. These threats could arise from natural causes or from manmade ones, or from the interplay of one with the other, as in the case of environmental degradation or global warming. Insofar as the targets of such threats (arising from a lack of maritime security) are concerned, these could be individuals themselves – or 'groupings' of individuals, such as societies and/or nation-states. When these threats address the regional fabric itself, the nation-states find themselves increasingly enmeshed in a complex web of security interdependence, which tends to be regionally focussed and thus a robust regional initiative ought to be a logical outcome of this regional focus.

For India, therefore, the concept of holistic maritime security incorporates both national and regional (Indo-Pacific) facets. At a regional level, it is these very human security issues that have been mentioned above that constitute the common interests. It is a common regional interest to create and consolidate a region in which the comity of nations is both intrinsic and assured...where every nation, big or small, is treated as an equal... where multiple options of governance are recognised as being functions of the independent choice of the people of each nation-state...where the people of every state of the region can live in dignity and peace...where poverty stands banished and prosperity sits in its place...where the state protects the individual and the individual preserves the state in a symbiotic relationship designed to establish and spread stability across the region... where malevolent non-State entities find neither spatial nor temporal room for manoeuvre... in sum, then our common interests are the absence-of or freedom-from threats.

Legislation

Security, as a function of maritime administration, is closely allied to legislation and there is much streamlining required to be done by India. India signed the UN Convention on the Law of the Sea (UNCLOS) on December 10, 1982 and formally ratified it on June 29, 1995. However, on August 25, 1976 itself, well before the formal conclusion of the UNCLOS deliberations, India enacted the "Territorial Waters, Continental Shelf, Exclusive Economic Zone and Other Maritime Zones Act, 1976". However, India also proclaimed its right to exercise sovereign legal powers relevant not only to customs and other fiscal matters but also "any enactment for the time being in force in India or any part thereof" in relation to 'specifically designated' areas within the Continental Shelf and the Exclusive Economic Zone and stated that "any enactment so extended shall have effect as if the continental shelf or the part thereof to which it has been extended is a part of the territory of India", retaining its right to "make such provisions as it may consider necessary for facilitating the enforcement of such enactment".

Since the International Treaty Law that has been signed and ratified takes precedence over national legislation and since, in the event of any contradiction between the two, national legislation must be amended to uphold the international obligations that the nation has committed itself to, there is, quite clearly, considerable legal work to be done to align the provisions of the "Territorial Waters, Continental Shelf, Exclusive Economic Zone and Other Maritime Zones Act, 1976" with the provisions, obligations, freedoms and restrictions that are contained in the subsequent UNCLOS, 1982. The Indian failure to do so has inevitably led to situations (such as the *Enrica Lexie* case involving two Italian Marines) where the Indian courts, being national instruments, have quite correctly ruled to uphold national law, but the validity of the un-amended national law, flying in the face of international treaty law ratified by India, has been questioned by the International Tribunal on the Law of the Sea (ITLOS).

Four International Models for Maritime Administration

In order to optimally streamline maritime administration, it would be useful to take a brief glance at the models adopted by other nations. Four models have been briefly touched upon – those of Australia, Canada, the European Union and the USA.

Australia[21]

The Australian Maritime Safety Authority (AMSA) is a statutory authority established under the Australian Maritime Safety Authority Act 1990 (the AMSA Act). As a Corporate Commonwealth Entity, it is also subject to the 'Public Governance, Performance and Accountability Act, 2013'. Its principal functions are:

(a) promoting maritime safety and protection of the marine environment
(b) preventing and combating ship-sourced pollution in the marine environment
(c) providing infrastructure to support safety of navigation in Australian waters, and
(d) providing a national Search and Rescue service to the maritime and aviation sectors.

Canada

Transport Canada

The principal organisation in Canada's maritime administration is 'Transport Canada'.[22] It promotes safe, secure, efficient and environmentally-responsible transportation of all kinds, including all forms of maritime transport and their nodes.

(a) ***Ports.*** While Transport Canada itself owns and manages several public ports, it also
   (i) Oversees Canada's 18 Canada Port Authorities (CPAs),
   (ii) Sets and monitors port and marine-facility service-standards,
   (iii) Sets and collects public port fees,
   (iv) Divests regional/local and remote ports under the Port Divestiture Program, and
   (v) Coordinates public notices and consultations

(b) ***Shipping.*** Transport Canada ensures compliance of all shipping with Complying with the Canada Shipping Act (2001) and it is the national authority for registration, inspection and certification of all commercial marine vessels.

(c) ***Marine Safety.*** Transport Canada is the authority for marine safety, including
   (i) Setting standards for vessel crew, design, equipment and use
   (ii) Monitoring and compliance of Canadian and foreign vessels
   (iii) Licensing and identification-marking and operator-competence evaluation of all pleasure craft
   (iv) Inspection of dangerous goods containers
   (v) Navigability of Canada's waterways
   (vi) Cargo-handling and ship-port interfaces
   (vii) Navigation and radio communications
   (viii) Environmental protection and response systems

(d) ***Marine Security.*** Transport Canada is responsible for developing and enforcing civilian transportation security regulations, including the Integrated Ship and Port Safety (ISPS) Code. The Canada Port Authorities, which own and operate Canada's major ports, as well as the private sector owners and operators, must ensure the security of their facilities under Transport Canada's regulatory requirements. Transport Canada additionally coordinates the Government of Canada's overall marine security policy with other federal departments and agencies such as the Department of National Defence (particularly the Navy) and the Canadian Coast Guard.

The Canadian Coast Guard

Naval operations in Canada's maritime environment are exclusively the responsibility of the Royal Canadian Navy. As such, the Canadian Coast Guard has no naval or law enforcement responsibilities. It is responsible for the following:[23]

(i) Search and Rescue,
(ii) Marine Communications and Traffic Services (including the Vessel Traffic-Management System [VTMS], the Automatic Identification System [AIS], and Communications Control System [CCS]),
(iii) Navigation aids,
(iv) Icebreaking, and
(v) Marine pollution-response.

Europe

The maritime administration of the European Union is handled by a triumvirate of organisations.

(a) The European Maritime Safety Agency (EMSA), one of the EU's decentralised agencies, is based in Lisbon, Portugal. It provides technical assistance and support to the European Commission and Member States in the development and implementation of the EU legislation on maritime safety, pollution by ships and maritime security. It has also been given operational tasks in the field of oil pollution response, vessel monitoring and in the Long Range Identification and Tracking (LRIT) of vessels.[24]
(b) On the other hand, maritime administration in Europe vis-à-vis fisheries is effected through the European Fisheries Control Agency (EFCA), headquartered in Vigo, Spain.[25] This is a European Union body established in 2005 to organise the operational coordination of fisheries control and inspection activities by Member States and to assist them to cooperate so as to comply with the rules of the Common EU Fisheries Policy in order to ensure its effective and uniform application. It makes a significant contribution to a Europe-wide level playing field for the fishing industry so that the European obligations are observed by everyone, and everyone in the sector is treated equally, wherever they might be operating. It also contributes

meaningfully towards sustainable fisheries by enhancing compliance with existing conservation and management measures to the benefit of the present and future generations. The EFCA, in cooperation with the European Border and Coast Guard Agency (FRONTEX) and the EMSA, also supports national authorities in carrying out Coast Guard functions.

(c) The FRONTEX performs the following roles:[26]

  (i) Monitors migratory flows and carries out risk analysis
  (ii) Monitors the management of external borders of the EU
  (iii) Provides operational and technical assistance to Member States
  (iv) Supports Search and Rescue (SAR) operations, and
  (v) Provides support to national Coast Guard authorities

USA

It is important to note that in the USA, too, the expression 'maritime administration' per se is limited in application to the US mercantile shipping, while the responsibilities and tasks involved in administering the vast maritime spaces of the USA are almost exclusively those of the US Coast Guard.

(a) The Maritime Administration Division (MARAD)[27] is a division of the Department of Transportation, whose main functions are to foster and promote the development and maintenance of an adequate, well-balanced United States merchant marine, sufficient to carry the nation's domestic waterborne commerce and a substantial portion of its waterborne foreign commerce, and capable of service as a naval and military auxiliary in time of war or national emergency. Towards fostering a robust manpower base, it vigorously supports maritime training and education. MARAD also seeks to ensure that the country maintains adequate shipbuilding and repair services, efficient ports, effective inter-modal water and land transportation systems, and reserve shipping capacity for use in time of national emergency.

(b) The US Coast Guard (USCG)[28] is that nation's oldest continuous seagoing service. Unlike in most other countries, the USCG is an integral branch of the military. It has 11 official missions, viz.,

  (i) Port & Waterway Security
  (ii) Search & Rescue (SAR)
  (iii) Aids to Navigation (ATN)
  (iv) Drug Interdiction

(v) Living Marine Resources
(vi) Marine Safety
(vii) Military/Defence Readiness
(viii) Migrant Interdiction
(ix) Marine Environmental Protection
(x) Ice Operations, and
(xi) Law Enforcement

While there is much to be learnt from a detailed study the organisational structures of these four global models of maritime administration, as also from their enabling and supporting legislations, perhaps the one that comes closest to mimicking India's federal structure is the European model.

Streamlining India's Maritime Administration

India needs a maritime administration that will undertake at least the following 21 Tasks in the 21st Century:[29]

(i) Proactively create and promote maritime awareness, maritime education and maritime public relations.
(ii) Ensure that the time-specific goals of the UN's Sustained Development Goal 14 are proactively met.
(iii) Provide informed professional inputs for the formulation of and amendments to the maritime legislation and its conformance with the international maritime law and the international maritime obligations committed to by India.
(iv) Enforcement of relevant laws and regulations within the Maritime Zones of India.
(v) Assurance of stable domestic and international mercantile shipping, including passenger-shipping.
(vi) Formulate and promote measures for the development of Coastal Shipping.
(vii) Formulate, promote and proactively advocate measures to increase the percentage of the Indian merchandise trade being carried in Indian-flagged vessels.
(viii) Proactively support and contribute towards the success of the SAGARMALA Project.
(ix) Undertake periodic and regular analysis of the functional efficiency of ports and harbour/port-facilities and formulate and pursue remedial/improvement measures.

(x) Exercise 'Flag State', Port State' and 'Coastal State' controls, inspections and responsibilities.

(xi) Provide, maintain and sustain requisite mechanical, electrical, electronic, sonic, optical and optronic aids to navigation, including lighthouses, lightships and buoyage.

(xii) Ensure the provision of maritime Search and Rescue and the promotion of maritime safety.

(xiii) Formulate and promote measures for the sustainable and environmentally sensitive exploration and exploitation of the living and non-living resources within and beyond the Maritime Zones of India.

(xiv) Formulate and promote measures for sustainable and ecologically sound fishing within and beyond the Maritime Zones of India, including the deconfliction of fishing, shipping and offshore economic development areas/zones.

(xv) Formulate and promote measures for the nurturing and development of human resources – in adequate quantity and quality – for the fishing, mercantile shipping, port, and shipbuilding industries.

(xvi) Ensure the enforcement of IMO-approved pollution-control measures for shipping.

(xvii) Ensure the enforcement of preventive, curative and punitive measures to assure protection of the marine environment.

(xviii) Promote measures to ensure maritime safety and the promotion of maritime security.

(xix) Promote the creation of multiple but coordinated means and channels of maritime communications with and between disparate maritime stakeholders.

(xx) Vigorously promote and develop maritime thought and concepts through proactive and sustained support to maritime think tanks and universities.

(xxi) Encourage, support and advocate regional approaches to better maritime administration such as the Indian Ocean Naval Symposium (IONS), the Indian Ocean Regional Association (IORA), the Western Pacific Naval Symposium (WPNS), Security and Growth for All in the Region (SAGAR), etc.

Organisational Issues

In contemporary India, given the largely incoherent nature of maritime administration, there is a need to develop a 'single-point-of-advice' solution, where one 'Maritime Adviser' would be the point of not merely 'coordination' but coherent 'integration' for the entire gamut of activities involved. Indeed, there appears to be a strong need for such a long-serving (if not permanent) 'Maritime Adviser' who has no other simultaneous functional responsibilities to discharge.

A strong case can be made that the Chief of the Naval Staff, by virtue of his ex-officio responsibilities, is, indeed, the actual 'maritime-security adviser' to the Government of India. The problem, of course, is that he is already dual-hatted as the commander-in-chief of the Indian Navy and the single-point (staff) adviser for military maritime security to the President of India through the government of the day. Moreover, as has been pointed out earlier, military maritime security is but one facet of maritime security (albeit a very important one) and, moreover, security per se is not the sole pursuit in maritime administration. There are, for example, critical issues of national development, legislation, environmental sustenance, the development and fostering of mercantile shipping, maritime-disaster warning and response mechanisms, etc., some or all of which are quite central to the business of maritime administration, even though they may not be as central to the business of maritime security. As such, an incumbent Chief of the Naval Staff, howsoever competent, has neither the training, nor the exposure, nor, indeed, during the period of his incumbency, either the time or the organisational and human wherewithal, to function effectively as the coordinator – and, even more difficult, the 'integrator' – of the many disparate stakeholders mentioned earlier in the paper. Moreover, continuity – an essential ingredient of experience-based expertise – is unavailable as any given Chief's tenure does not exceed three years. All this is, of course, equally true of other high functionaries such as the Director General Coast Guard, the Director General Shipping, etc. All of them have the same infirmities during the period of their respective incumbencies. On the other hand, to choose a generalist (an IAS official, for example) who may well have much experience of general administration, but who by the nature of his own training and exposure has little or no experience or expertise in the multifarious facets of the maritime domain, would be an even greater disaster.

If incumbency-based responsibilities are the main hindrance, a retired (and hence greatly experienced) high-functionary from one of the subsets of the maritime domain can provide the requisite 'single-point' advice. In this context two issues merit attention.

The first concerns the manner in which successive governments strike a balance between competence and seniority while selecting/appointing or approving the appointment of any given high-functionary. There is sufficient evidence of great controversy in the striking of this balance and so it is by no means necessary that a given 'Chief' or 'Director General' is actually the most competent amongst his peers or immediate juniors. Thus, when he retires, he is very unlikely to be any more competent than when he was serving and had a far greater number and variety of advisers and sub-domain experts to advise and guide him, Perhaps this could be resolved by having a minimum criteria for balance between the responsibilities of various ranks/positions and jobs/assignments/appointments tenanted. An illustrative example (though by no means either an exhaustive or a limiting one) might be to specify two-star Flag Officers and equivalents as a minimum eligibility criterion. However, given the profusion – and hence dilution – of stars in services other than the military, this will need greater specificity.

The other problem is that the governments engaged in the process of either selection or confirmation of an apparently or potentially powerful appointment/position often covertly seek pliability even while overtly and very strenuously rejecting this trait. This leads to the question of how (and whether) the government of the day could be removed from the equation altogether, or, have its influence reduced to a minimum. The creation of a constitutional authority (on the lines of the Chief Election Commissioner), might be the answer here, but cannot, in and of itself, provide a guarantee of unimpeachable character and widely respected competence. The dilution of all our reporting mechanisms (typically Confidential Reports) has been marked and, coupled with the ongoing tendency to equate the positions of different professions/services in solely monetary terms, makes for a problem that defies easy solution.

A related problem is one of providing authority commensurate with the enormity of the responsibility being given. A failure to provide authority sufficient in scope, range and delivery to overcome stovepiped turf-based loyalties and ensure active involvement and participation at the appropriate level has been the bane of many organisational structures in India. As soon as the absence of manifest authority (power) is discerned, all such organisational structures immediately become toothless and ineffective. They are, more often than not, attended by the lowest-viable (and frequently the least-wanted) functionary from each organisation, other than the one that the chairperson is from. This has been the uniform experience of the Joint

Operation Centres set up after 26/11 and is seen in several other multi-disciplinary or multi-organisational committees (such as the Offshore Security Coordination Committee, the Maritime States Development Council, etc.) the Indian institutions and individuals tend, dichotomously, to respect power but, wherever possible, to disregard, demean and deride authority, especially if such authority is seen to be powerless in its own specific domain-hierarchy. Thus, the provision of adequate and manifest power and authority is a *sine qua non* for the streamlining of maritime (multidisciplinary and multi-agency) administration.

Another option is to reduce the absolute power of the individual concerned by creating a hierarchical set of boards – a multi-layered committee system – along the lines of the National SAR Board or the Offshore Oil Security Committee. This might solve some of the foregoing problems, but the risk of the tongue-in-cheek definition of a committee being twelve individuals who, individually, can do nothing, but who, collectively, decide that nothing can be done, must be taken into account as well. This brief frivolity notwithstanding, the support for a hierarchy based upon Boards/ Committees is provided by the realisation that given the sheer width, range and scope of responsibilities, and the tasks and activities involved in maritime administration, it is difficult to believe that a classic pyramidal hierarchic structure with one individual at the apex functioning as a single-point-of-advice to the government will work. A further weight in favour of this option is provided by all the four foreign models considered.

In 2014, there was much speculation over the seemingly imminent setting up of the National Maritime Authority (NMA) as an apex federal body by the current government. This is a widened conceptualisation of a similar concept – dealing with the various aspects of maritime security. Even the 2001 GoM report on reforming the national security system had underlined the need for "an apex body for management of maritime affairs for institutionalised linkages among the Navy, Coast Guard and ministries of central and state governments". The desirability of this was very strongly reiterated in the wake of the terror attacks in Mumbai between 24 and 26 November 2008 (the 26/11 attacks), and even figured on the agenda of the Cabinet Committee on Security in the shape of a Maritime Security Advisory Board (MSAB) with a Maritime Security Advisor as its chief. However, the government then in power had junked it due to turf-sensitivities in the office of the NSA and due to electoral compulsions. The 2014 concept was expected to move beyond security – certainly beyond military security – and the NMA was envisaged as a centralised agency overseeing the activities of the entire domain. This activity needs to be brought to fruition.

However, the streamlining of maritime administration will require more than the creation of an apex body with a Chief Maritime Adviser (NMA) at its head. A number of subordinate and ancillary organisational structures will also be required at the state level and within the country's littoral belt.

Legislative Issues

The need for a strong foundation of legislative support right from the inception can hardly be overemphasised. In determining an appropriate structure for this, the National Law University may well be a good place with which to begin. Without a sound legislation that is in consonance with the international treaty law, the maritime administration would remain weak if not capricious. However, even here, the degree of scholarship on International Maritime Law would need to be significantly enhanced, perhaps through the creation of a Chair for International Maritime Law. Adequate funding and support from the Government must be built into the approval mechanisms so as to permit the enrichment of domestic scholarship by an intensive and extensive interaction with legal entities – particularly from the European Union. It is reiterated that the multilingual, multi-ethnic EU, with its prickly sensitivities is the closest that we can get to in mimicking India's fractious federal structure. The aim must be to ensure that the Indian domestic law is brought into consonance with such international laws as India has ratified. In fact, the process of signing and ratification by India of international treaties and conventions need to be examined in detail, itself streamlined, and, formally taught at various fora of executive functioning. Within the Navy, the Coast Guard and the Directorate General of Shipping, there is a specific need to have law officials (Judge Advocates in the Navy and the Coast, for example) thorough with all aspects of the International Maritime Law, as also with the powers of the Parliament and the Executive in India to enter into treaties and the effects hereof.

Coordination Issues

A major task of coherent maritime administration is the coordination between the various stakeholders operating or interested in the maritime spaces relevant to India. This is where most organisational structures flounder. However, if the foregoing aspects, viz., organisation and legislative support are correctly, sincerely and carefully addressed, the coordination issues will be relatively simple. Once again, there appears to be a great deal of merit in drawing from the experiences of the European Union. The following schematic shown in Fig. 1, offers an indication of the coordination structure with intergovernmental and multinational agencies.

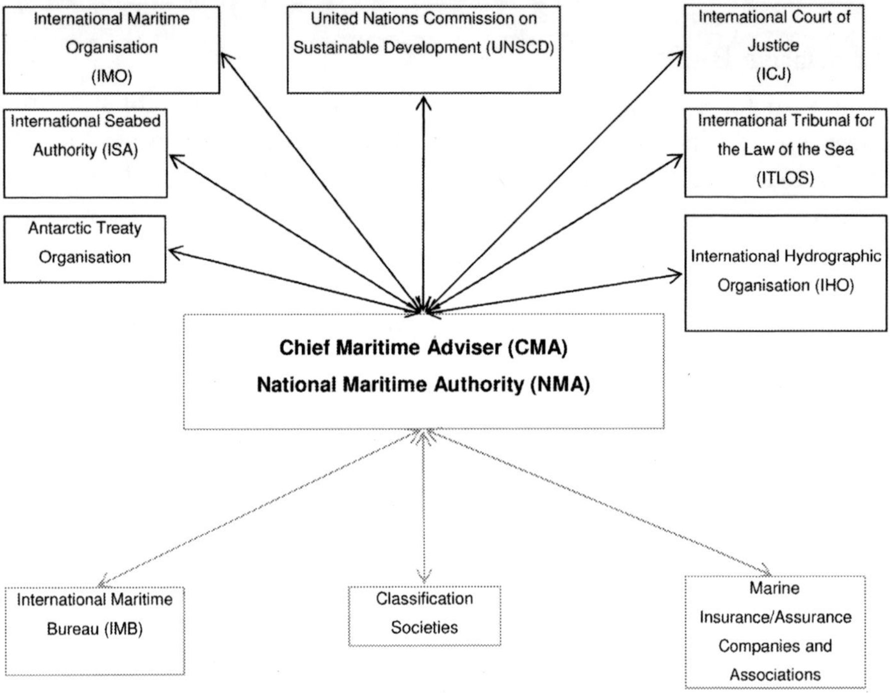

Figure 1. Interaction between NMA and other intergovernmental agencies

Insofar as the interfaces between the NMA and the national authorities is concerned, the schematic diagram as shown in Fig. 2, provides some idea of the coordination functions to be attained and sustained. It would be obvious that these coordinating structures would need to be repeated at the level of the maritime states and union territories.

Conclusion

It is self-evident that if India is to honour its commitments made in the UN General Assembly and specifically in respect of Goal 14 to meet its time-specific targets, India's maritime administration mechanisms will have to be

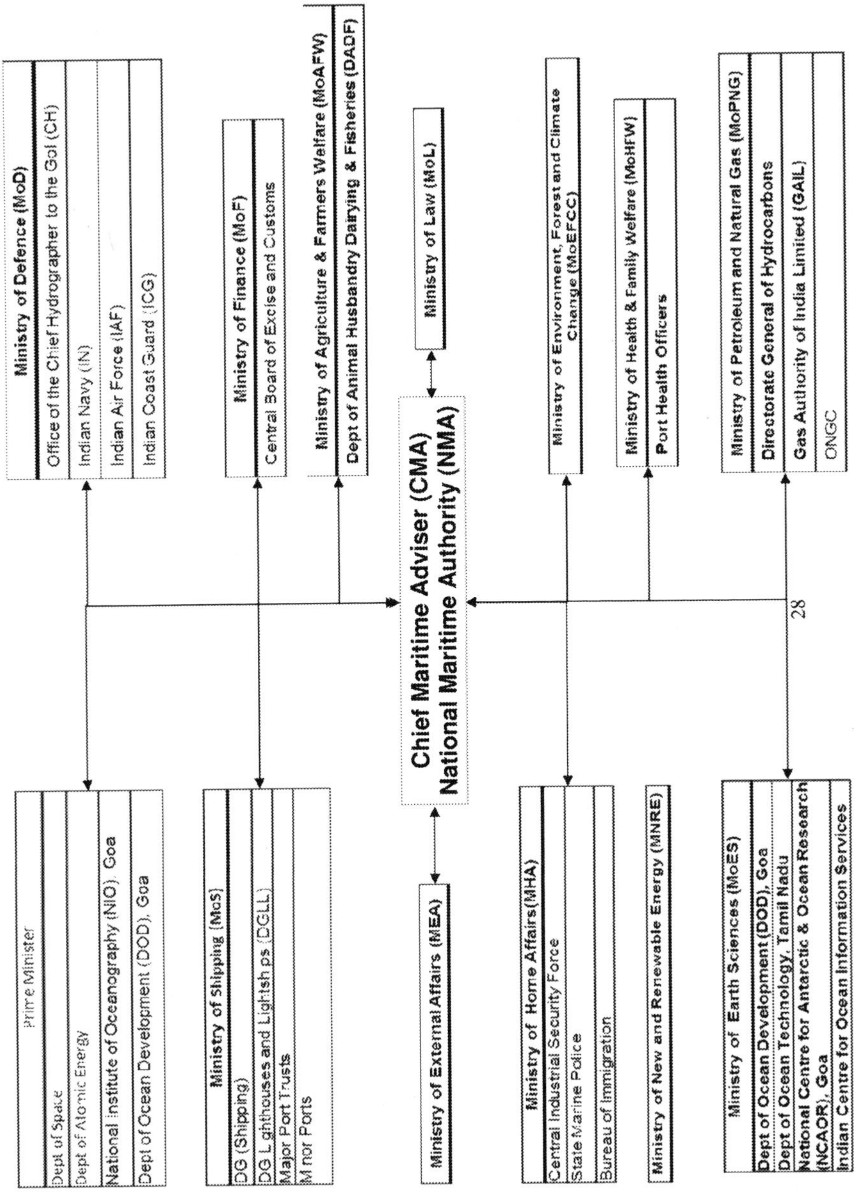

Figure 2. Coordination between the NMA and other ministries

pretty much reinvented. It is also vital that we avoid the trap of being sucked into thinking about maritime administration solely in terms of mercantile shipping, or, for that matter, solely in terms of maritime security either.

Streamlining the organisational structure requires the speedy creation of a central coordinating and controlling body, preferably a constitutional entity, which is suitably competent and has authority. India is barely six months away from receiving its first 'report card' from the HLPF at the UN. In the reinventing of our maritime administration structure and mechanisms, it is critical that India draws from the pool of international experience. Of the many models available (only a tiny sampling of which has been touched upon in this chapter), perhaps the one that is the most relevant to India, is the European Union model. Significantly, the European model has an extremely strong foundation in European law and this is an area that India must address or attack first. It is also the one from which India might learn the most.

Insofar as developing India's own organisational structure is concerned, this chapter has proposed a Board or Committee system. The chapter has also spelt out the risks inherent in adopting such an approach without adequate and evident empowerment, while providing some thoughts on how the Chief Maritime Adviser and the National Maritime Authority might best be insulated from the vagaries of short-term political or electoral considerations. The broad qualifications of the Chief Maritime Adviser have been touched upon, with strongly cautionary inputs being provided against choosing a tenure-limited serving senior-functionary.

Finally, it may have been noticed that the very first of the 21 tasks set-forth as minima for India's maritime administration in this 21st Century is to proactively create and promote maritime awareness, maritime education and maritime public relations. The pressing need for this cannot be emphasised strongly enough.

REFERENCES AND NOTES

1. "A New UN Forum to boost Sustainable Development", UN Department of Economic and Social Affairs, 24 September 2013, http://www.un.org/en/development/desa/news/sustainable/high-level-political-forum.html (last-accessed on 23 Jan 17)
2. "Transforming Our World: the 2030 Agenda for Sustainable Development", Resolution 70/1 of the UN General Assembly, 25 September 2015, https://sustainabledevelopment.un.org/post2015/transformingourworld (last-accessed on 23 Jan 17)
3. UN Sustainable Development Knowledge Platform, https://sustainabledevelopment.un.org/hlpf/2017 (last-accessed on 23 Jan 17)
4. "Our Oceans, Our Future: Partnering for the Implementation of SDG 14", The UN Ocean

Conference, 05-09 June 2017, New York, https://sustainabledevelopment.un.org/oceanconference/ (last-accessed on 23 Jan 17)

5. "Introduction to IMO", IMO website, http://www.imo.org/en/About/Pages/Default.aspx (last-accessed on 24 January 2017)
6. International Maritime Bureau, https://icc-ccs.org/icc/imb
7. World Trade Statistics, 2016, p. 35, https://www.wto.org/english/res_e/statis_e/wts2016_e/wts2016_e.pdf
8. Union Budget (India), Chapter 4: External Sector, p. 65, http://indiabudget.nic.in/es2015-16/echapvol2-04.pdf
9. Knoema World GDP Ranking, 2016, *op cit.*
10. World Trade Statistical Review 2016, Tables A4 and A5, pp. 92-93, https://www.wto.org/english/res_e/statis_e/wts2016_e/wts2016_e.pdf
11. Government of India Business Knowledge Resource Online (Business.gov.in) National Level Infrastructure: Maritime Transport, http://www.archive.india.gov.in/business/infrastructure/maritime_transport.php
12. "Indian Shipping Statistics 2014". Government of India, Ministry of Road Transport and Highways (Transport Research Wing), New Delhi, p. i, http://shipping.nic.in/showfile.php?lid=1960 (last-accessed on 23 January 17). Specific Tables are also available https://data.gov.in/catalog/indianfleet-statistics-0#web_catalog_tabs_block_10
13. "The Black-Grey-White Lists (Valid 1 July 2016 - 30 June 2017)", http://www.tokyo-mou.org/doc/Flag (accessed on 21 January 2017)
14. "Report of the Expert Committee Constituted for Comprehensive Review of the Deep Sea Fishing Policy and Guidelines", Department of Animal Husbandry, Dairying & Fisheries, Ministry of Agriculture, Government of India, August 2014, p.9, http://www.sherrylegal.com/articles/7/meenakumari committee report fishing.pdf
15. Marine Products Exports (Market Wise) 2015-16, Marine Products Export Development Authority (MPEDA) website, http://mpeda.gov.in/MPEDA/marine_products_exports.php#
16. "In Pursuit of a $500 Million Harvest, India is Gearing up for a Deep Dive", Quartz India, https://qz.com/282373/in-pursuit-of-a-500-million-harvest-india-is-gearing-up-for-a-deep-dive/
17. "Tuna Advisory", Indian National Centre for Ocean Information Services INCOIS), 02 December 2013, http://www.incois.gov.in/MarineFisheries/TunaAdvisory (accessed on 24 January 2017)
18. "Report of the Expert Committee Constituted for Comprehensive Review of the Deep Sea Fishing Policy and Guidelines", *Ibid*, p.30.
19. "Transforming Our World: the 2030 Agenda for Sustainable Development", Resolution 70/1 of the UN General Assembly, 25 September 2015, p. 9, https://sustainabledevelopment.un.org/post2015/transformingourworld (last-accessed on 23 Jan 17).
20. Address by Dr Manmohan Singh, erstwhile Prime Minister of India, New Delhi, 14 February, 2008, http://archivepmo.nic.in/drmanmohansingh/speech-details.php?nodeid=633 (last-accessed on 24 Jan 17)
21. AMSA Website, https://www.amsa.gov.au/about-amsa/
22. Transport Canada Website, http://www.tc.gc.ca/eng/menu.htm
23. Canadian Coast Guard website, http://www.ccg-gcc.gc.ca/eng/CCG/Home
24. EMSA website, http://www.emsa.europa.eu/about.html
25. EFCA website, http://www.efca.europa.eu/en/content/objectives-and-strategy
26. FRONTEX website, http://frontex.europa.eu/about-frontex/mission-and-tasks/

27. MARAD website, https://www.marad.dot.gov/about-us/maritime-administration-at-a-glance/
28. US Coast Guard website, https://www.gocoastguard.com/about-the-coast-guard/discover-our-roles-missions
29. "Report on Maritime Affairs (Summary)", Maritime Bureau, Ministry of Land, Infrastructure, Transport and Tourism, Government of Japan, July 2008. www.mlit.go.jp/common/000118685.pdf

11

Planning and Financing Blue Economy

P. Sekhar

The Blue Economy is a novel concept which is envisaged as the integration of marine and inland water-based economic development with the principles of social inclusion, environmental sustainability and innovative, dynamic business models. India has naturally been a maritime country with more than 7,500 km of coastline, above 2 million sq. km. of Exclusive Economic Zone (EEZ), more than 1,200 islands and 195,095 km length of river and canal network to leverage this natural advantage. It is estimated that the value of the Blue Economy potential in India is almost US$ 1 trillion. It is important to plan for appropriate support and to create a sustainable business environment for onshore businesses and for sectors related to the Blue Economy. Secured governance offers a strategy for the government to utilise value in a systematic way that leads to minimum government investment and public expenditure for infrastructure development. The Indian government must promote long-term investments by the private sector which generates strong returns and ecosystem benefits and needs to play a regulatory and supervisory role. Moreover, foreign investors are also ready to support the marine infrastructure development as the most effective way to achieve sustainable growth. The final goal of the Blue Economy is not just economic growth and quality of life for all people, but to meet the needs of the present and future generations with respect to economic, social, environmental as well as cultural aspects.

Introduction

The Blue Economy is a novel concept which envisages the integration of marine and inland water-based economic development with the principles of social inclusion, environmental sustainability and innovative, dynamic business models. The Indian waters have a huge potential to deliver growth and jobs and to generate a gross added value of almost US$ 1 trillion in the coming years.

India's Blue Economy at a Glance

India is endowed with a rich and vast diversity of natural water bodies, seas, rivers and canals; reservoirs; dams, tanks, lakes & ponds; lakes and derelict water bodies; and brackish water.[1] .There are in total, 9 states and 4 Union Territories along the coastal regions. The marine and inland water resources of India are briefly shown in Table 1.

Table 1. Marine and Inland Water Resources of India

Coastal length (in km.)	7,516.6
Exclusive Economic Zone (EEZ) million sq. km.	2.37
Islands (in nos.)	1,208
Rivers & Canals (length in km.)	195,095
Other Water Bodies (area in Mha)	
Dams (in nos.)	4,857
Reservoirs	2.93
Tanks & Ponds	2.43
Flood Plain Lakes & Derelict Water Bodies	0.80
Brackish Water	7.31

The Indian Ocean is one of the leading maritime trade and commerce zones in the world. It is also gifted with vast unexploited natural resources and is expected to yield a number of benefits for India. Maritime transport handles 78 per cent of the world trade by volume and over 70 per cent by value. India's seaborne trade is 90 per cent by volume and 70 per cent by value. Hence, a conscientious activity is required in the coastal region to promote maritime trade so that India gets a larger share in terms of its own ships carrying its trade. Apart from India's strategy to economically develop the coastal areas, the Indian Navy has also mapped a long-term plan for expanding its influence area in the Indian Ocean and beyond.

Key Industries of the Blue Economy

The marine and inland water resources are rich with energy and food sources and provide a gateway for many industries from transport to maritime security, tourism to oil and gas for operation.

Smart Port City

The smart port city is a recent concept in town planning developed along the seacoasts that combine the latest technology in an urban and environmental context, thereby creating a port city that is increasingly livable and at the same time flourishes economically. The Indian coastline is serviced by 12 major ports and 200 notified minor and intermediate ports.

Marine Tourism

India has immense potential in the marine tourism segment. Each coastal state has its own seafood cuisine, relaxing spas, diving and water sports and great places to stay for a holiday.

Marine Transportation

Sea arrivals constituted about 0.4 percent[2] of India's total foreign arrivals in year 2014. It is estimated that by the year 2030–31, the Indian cruise tourism market will attract 1.2 million "tourists".[3] Even the cargo traffic through seaways, which recorded 1,052 Million Metric Tonnes (MMT) in 2015, was expected to reach 1,758 MMT by 2017.[4]

Ship and Boat-Building Industry

In India, although private players hold the majority of the shipbuilding capacity, three public entities have the required infrastructure and capacity to build large vessels. As on 31 March 2015, the Indian shipyards had an order book of 258 ships with Deadweight Tonnage (DWT) of 2,692.70 thousand tonnes.[5]

Fisheries

India's marine fish production has varied during the last four years with a reported catch of 3.32 million tonnes in 2012–13, 3.44 million tonnes in 2013–14, 3.66 million tonnes (provisional) in 2014–15 and 3.58 million tonnes (provisional) in 2015–16, respectively.[6]

Offshore Oil and Gas

In 2015, the oil production in the country reached 0.75 mbpd as compared to 0.76 mbpd in 2014, registering a decline of 0.85 per cent. In 2014, there were 5.7 billion barrels of proven oil reserves and 1.4 tcm of gas proven reserves and produced 33.66 bcm of gas in 2015 which was expected to rise and reach 33.73 bcm in 2016.[7]

Ocean Renewable Energy

Oceans can produce two types of energy: thermal energy from the sun's heat and mechanical energy from the tides and waves. The economically viable tidal power potential identified in India is of the order of 8,000–9,000 MW with about 7,000 MW in the Gulf of Cambay, about 1,200 MW in the Gulf of Kutch and less than 100 MW in the Sundarbans. The total potential along the 6,000 km of coast is about 40,000 MW.[8]

Marine Biotech Industry

The biotech industry can be broadly divided into the following segments – Biopharmaceuticals, Bio-industrial and Bioinformatics. All of them have a large potential in India.

Deep Sea Mining

The deep sea hosts a much diverse and unique ecosystem than the terrestrial ecology. With the license for exploration, India will start looking for polymetallic sulphides that are rich in copper, zinc, gold and silver in the Indian Ocean basin.

The Blue Economy Concept – Secured Governance Perspective

"The Blue Economy is envisaged as the integration of water based economy including inland and marine based sustainable economic development with the principles of social inclusion, environmental sustainability and innovative, dynamic business models".

Secured governance presents a vision of rapid economic development based on natural resource utilisation and urban expansion through the next five years. It prioritises the immediate development needs and the interests of coastal developers and extractive industries over the proactive preservation of ecosystem services. As the scenario was developed over time, it came to represent possible construction and development that could occur with minimal government investment. Substantial benefits are to be derived through this type of secured governance functions via many interacting mechanisms, which define and control them.

Apart from the obvious benefits of additional revenue through coastal development, one major advantage is the improvement in exports and imports for India. Refining valuation methods are also going to lead to more rational coastal land-use decisions, and would improve the understanding on the ecological–economic mechanisms that contribute to the value of natural capital assets.

Secured governance offers a strategy for the government to get all basic infrastructure development with a negligible investment. It is a concept of developing techno-economic corridors and connecting hubs which will act as growth centres for individual sectors. The very concept of "Secured" here implies a secured convergence or knitting with various sectors of the economy.

A healthy marine ecosystem provides an irreplaceable foundation for fisheries, recreation and tourism industries, as well as less recognised services such as waste processing, hydrological cycling, storm protection and nutrient cycling – all of which contribute to the habitability of a place, the region and the planet as a whole. The innovative Blue Economy is driven by business-level innovation using locally sourced resources, with a focus on job creation, building social capital and generating multiple cash flows by stimulating entrepreneurship and business-model innovation.

Historically, there has been a close correlation between economic growth and environmental degradation. As communities grow, environment declines. Hence there was an emergence of various policy tools which encouraged sustainable production and consumption patterns. In addition, frequent dialogue, knowledge management, technology transfer, education, training and capacity building must be developed so that sustainable industrial practices can spread throughout the world. Coastal plains are a geologic feature found on both the eastern and western coasts of India. The coastal area of India has the potential to generate multiple revenue streams.

Blue Economy Planning

Blue Economy Planning has many other similar associated terms such as integrated coastal zone management, coastal and marine spatial planning, maritime planning and coastal development planning. This requires planning, managing and effectively governing the use of marine space and resources, applying inclusive methods and the ecosystem approach. All relevant uses of marine space and resources must be accounted, planned, managed and governed through forward-looking, precautionary, adaptive and integrated processes that ensure the long-term health and sustainable use of the sea, while also taking into account human activities on land.

Such processes must be participatory, accountable, transparent, equitable and inclusive, in order to be responsive to the present and future human uses and needs, including the needs of minorities and the most vulnerable groups in society.

Sustainable Marine Fisheries

"While keeping sustainability of the resources at the core of all actions, the policy framework will meet the national, social and economic goals and wellbeing of the fishing society and is aimed to channelize the coordination and management of marine fisheries in the country during the next ten years".

Even if considering the wide confidence intervals of catch and potential yield estimates, there is a need for taking a precautionary approach in line with the global standards regarding marine fish harvests. With regard to inshore waters, the Government lays emphasis on maintaining the harvest at about the present levels of Maximum Sustainable Yield (MSY), with sustainability and equity as the key principles. It will also be ensured that the small-scale character of the entire marine fisheries industry is retained. The management of fisheries will follow an integrated approach, blending traditional knowledge and science with commercial and industrial principles and more effective engagement of both primary stakeholders, as also those engaged in ancillary activities to ensure that the fisheries are ecologically and economically sustainable.

Knowledge management will be an approach to foster quick and easy availability of information on key attributes of marine fisheries sector, such as resource abundance and distribution; real-time resource maps; productivity assessments; real-time Potential Fishing Zone (PFZ) advisories and weather forecasts for the benefit of fishermen. The use of Information Technology (IT) and space technology will be put to optimum use for harnessing the benefits in support of the fishing community. Spatial and temporal closures have assisted in sustaining the marine fish wealth of the country.

Sustainable Coastal Tourism

Coastal tourism is a cognitive operation involving tourists and the people and the places they visit for entertainment, particularly the coastal surroundings and tehir natural and cultural resources. Coastal tourism destinations fall along an urban–rural continuum. Sun, beaches, and warm ocean waters have become standard vacation requirements for many tourists.

From an environmental perspective, tourism development can lead to degradation of ecosystems. This, of course, is unavoidable with the building of airports, ports, road systems, hotels, resorts and other facilities. Economic growth in the coastal region brings the risk of environmental damage, as it puts an increased pressure on the environmental resources. But policymakers driven by the concept of sustainable development will necessarily work to assure that the growing economies remain securely attached to their ecological roots and that these roots are protected and fostered so that they may support growth over the long-term.

Integrated Coastal Zone Management of the Beaches

Coastal zone management helps develop approaches and strategies that allow for better planning and management of tourism activities in coastal zones. This is not an easy task as these are highly fragile environments and are targets for many other human-induced development activities. Integrated Coastal Zone Management (ICZM) has been recognised by many tourism operators and policymakers as a way for the sustainable development of coastal tourism.

A better dissemination of the existing information should be achieved. For that, an effective interaction of the existing government bodies that interact with the coastal management is necessary. Besides, the information should be accessible to everyone. With regard to the latter aspect, scientific journalism has to be kept safe and favoured, as well as conferences where all the agents related with the coast can exchange their ideas, results or methodologies. The improvement of environmental education is important for the sustainable development of the coast.

Long-Term Benefits

There are clear long-term benefits when the private sector sets up and manages small marine parks such as in resource protection, environmental awareness and economics. Over-fished and depleted reefs adjacent to and upstream of the marine park are being restocked and local people and tourists are educated about related issues. Private management is considerably less costly and more efficient than government-controlled management bodies set up by over-funded donor projects.

Transferability

Well-managed marine parks and coral reefs are attractive for marine tourism and generate considerably more income than fisheries and other resource extraction activities. But these require a long-term security of tenure and also conservation-minded investors.

Sustainable Maritime Transportation

The maritime transportation system is part of a larger logistics chain that services the cross-border distribution of goods. It is subject to a complex array of policies and regulations. In order to provide a seamless and reliable service in an efficient manner, the maritime transportation system must deliver safe, secure, efficient and reliable transport of goods across the world, while minimising pollution, maximising energy efficiency and ensuring resource conservation. To achieve this, the complexity of the interrelation among the actors in the maritime transportation system should be recognised and taken into account when addressing specific actions. A sustainable maritime transportation system requires a well-organised administration that cooperates internationally and promotes compliance with global standards, supported by institutions with relevant technical expertise, such as classification societies acting as recognised organisations (i.e. organisations entrusted by a flag State to carry out mandatory inspections and surveys on its behalf).

In order to operate with high efficiency, a sustainable maritime transportation system requires coordinated support from the shore-side entities intrinsic to shipping, such as providers of aids to navigation, oceanographic, hydrographic and meteorological services, search and rescue services, incident and emergency responders, port facilities, trade facilitation measures and cargo-handling and logistics systems.

A sustainable maritime transportation system will need the collaboration of shore-side actors –from both the industry and the Government – for the protection of seafarers, in order to ensure that the system's social integrity does not become eroded and that qualified, professional seafarers have an attractive work environment.

Besides, the shipping sector needs external assistance, such as from the Navy or shore-based action, to meet security needs. However, it must also take preventative measures to address security threats arising at sea or in ports, which endanger both cargo and crew.

Sustainable Shipping Industry

Shipping causes a wide range of harmful impacts on the marine environment.

Most of the damage comes from the release of polluting substances. Intentional and unintentional discharges of oil, chemical cargo residues, garbage and cleaning agents, anti-fouling paint, exhaust and other air emissions and non-indigenous species from ballast water have an adverse impact on the seas. Copious amounts of waste generated on board ships are still being thrown overboard, harming the marine environment, leaving beaches scattered with empty oil drums, jerry cans, strapping bands, industrial packaging, spray cans, plastics, paint brushes, metal scrap etc.

Sustainable Marine Renewable Energy

Marine renewables are set to play a key role in reducing carbon emissions, and help to reduce climate change impacts on biodiversity. Marine renewable energy can be extracted from three main sources: wind, wave and tide. In addition to these, there are also initial investigations into the feasibility of using marine algae to produce energy. A variety of different technologies are currently under development throughout the world to harness this energy in all its forms including waves (40,000 MW), tides (9,000 MW) and thermal gradients (1,80,000 MW). Deployment is currently limited but the sector has the potential to grow, fueling economic growth, reduction of carbon footprint and creating jobs not only along the coasts but also inland along its supply chains.[9]

Tidal Energy

It is produced by the rise and fall of tides from the gravitational influence of the sun and moon. It can take two forms: tidal current and tidal range. Tidal energy is scalable; it can start with one turbine and over time, scale up to multiple turbines. Tidal current energy is a renewable, zero-emission option with a high power density.

Wave Energy

It is a concentrated form of solar energy and ocean waves that forms as a result of the sun heating the ocean surface. The motion of the waves can be used to power turbines to generate electricity.

Sustainable Deep Sea Mining

The deep sea mining technology is still at an early stage, and large parts of engineering systems are or may be transferred from the oil and gas as well as dredging industries. For seabed mining, new types of sea floor tools need to be taken into consideration. A research on the viability of seabed exploitation demands comprehensive analyses of multidisciplinary determinants such as economics of resources and their exploitation, environmental effects, societal benefits and legal constraints. Fig. 1 shows the triple bottom line concept of sustainability.

Economic Dimension

"Economic prosperity involves the creation and distribution of goods and services that will help to raise the standard of living around the world. Open, competitive, international markets that encourage innovation, efficiency, and wealth creation are fundamental aspects of sustainable development".

Environmental Dimension

"Ecologically sustainable companies use only natural resources that are consumed at a rate below the natural reproduction, or at a rate below the development of substitutes. They do not cause emissions that accumulate in the environment at a rate beyond the capacity of the natural system to absorb and assimilate these emissions. Finally, they do not engage in activities that degrade ecosystem service".

Societal Dimension

"The social equity principle ensures that all members of society have equal access to resources and opportunities. Human needs not only include basic needs such as food, clothing, and shelter, but also include a good quality of life such as healthcare, education and political freedom".

Stimulating Innovation

Technology may show the way for persistent environmental problems. Stimulating innovation in the maritime sector is crucial, keeping in mind that a technical solution to one problem should not lead to others. There is a crosslinkage between innovation and environmental policy. Once technologies are available, this will allow for more ambitious policy instruments.

Education and Public Awareness

Environmental awareness of the crew is of vital importance to stimulate clean shipping. Educating stakeholders not directly involved in operating a ship also helps in implementation of cleaner environmental measures.

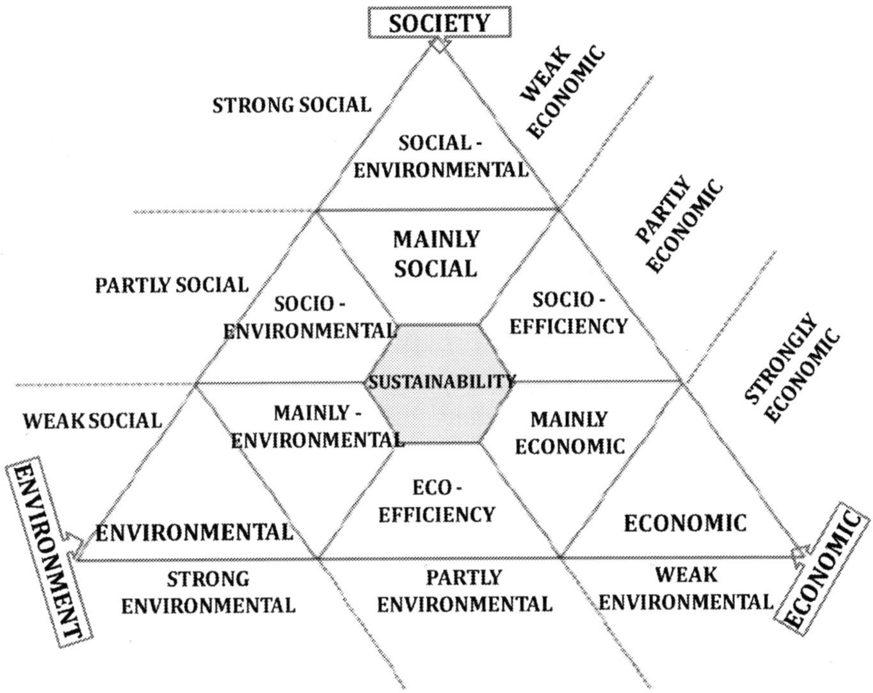

Figure 1. The Triple Bottom Line Concept of Sustainability[10]

Financial Instruments

Port authorities and governments can create strong incentives like lower berthing charges for clean and efficient ships and install fines for polluters, in line with the 'polluter pay principle'. Port dues together with environmental services like port reception facilities stimulate shipping companies to optimise their environmental performance.

Influencing Logistic Chains

Many actors are involved in distributing goods, from producer to consumer, with a wide range of stakeholders in between. Cargo owners, for instance, may demand environmental requirements of their ocean carriers that are more ambitious than what is provided through the current legislation.

Regulation

Existing rules are the bottom line for shipping, and new rules should make it more difficult for offenders to stay in business. Strengthened enforcement remains important if polluters do not get caught or if the fines are too low; environmental offences will remain a minor business risk. New regulations should encourage innovation, not only setting a standard for now, but also one for the near future.

Blue Economy and Financing

The Blue Economy is based on harnessing the ocean economy. The ocean economy includes industries explicitly tied to the ocean and industries partially related to the ocean and located within the shore-adjacent zone. Compared with the traditional Ocean Economy, the Blue Economy is more scientific and profound and covers a wider range of subjects. It attaches more importance to the development of high-level marine industries, a balanced utilisation of ocean and land resources and a scientific innovation as well as protection of the marine ecology.

The development of the Blue Economy depends on financial support. Financing plays an important role in developing the blue economic zone. Moreover, the development of a blue economic zone provides both opportunities and challenges for finance. An efficient implementation of this strategy depends on sufficient funds and reasonable resource allocation.

Currently, there is no effective financing system in the blue economic zone, which obstructs the development of the Blue Economy. The Sagarmala Coordination and Steering Committee (SCSC) will provide coordination between various ministries, state governments and agencies connected with the implementation and will review the progress of implementation of the National Perspective Plan, Detailed Master Plans and Projects. It will, inter alia, consider issues relating to the funding of projects and implementation. This committee will also examine the financing options available for the funding of projects and the possibility of public–private partnership in project financing/construction/ operation. All efforts would be made to implement those projects through the private sector and through the Public-Private Participation (PPP), wherever feasible. Funds envisaged for starting the implementation of projects in the initial phase of the Sagarmala Project were projected at Rs. 692 crores for the FY 2015–16. A further requirement of funds will be finalised after the completion of the Detailed Master Plan for Coastal Economic Zones for future years. These funds will be used for the implementation of projects by line ministries in accordance with approvals by the SCSC.[11]

Foreign funds strongly support the marine infrastructure development as the most effective way to achieve sustainable impact in the blue economic growth. In all these blue economy industries, foreign funds will continue to drive greater results and value for money through strategic partnerships with the private and public sector.

Current Financing Problems in the Blue Economic Zone

Commercial Banks

Banks can not satisfy the enterprise's needs for funds. Both demanders and suppliers of funds face risks and pressures.

Private Equity Funds

The blue economic zone lags behind other big cities in developing private equity funds. There are no policies to encourage private equity funds in the blue economic zone. Considering that the whole level of venture capital and private equity development is low in India, it is difficult for enterprises in the blue economic zone to get funds through these channels.

The Secured Governance Strategy in the Blue Economy Concept

One of the strategies of Secured Governance is to utilise the Blue Economy concept as a tool to implement smart city development in the islands for an economic growth of the nation. The final goals are not just economic growth and quality of life for all people, but sustainable smart cities while ensuring that it meets the needs of present and future generations with respect to economic, social, environmental, as well as cultural aspects. The smart cities in islands will help the Indian economy to provide long-term funding and more employment opportunities in the future. India has 1,208 individual islands in the smart city scheme and it is an ambitious economic development programme aimed at consolidating international business and financial hubs by creating ideal conditions for working, living and spurring investment. It provides technology-driven facilities to the business community and creates a vibrant city lifestyle.

Singapore's case study provides an interesting example of a nation's approach towards being a smart city. The small island nation sits just one-degree north of the equator, consistently hot and humid year round. The island lacks basic resources; it has no energy deposits, no forests and even no farms. However, despite lacking natural resources and hinterland, Singapore has grown into a global commerce centre and its economy is based on external trade and its skilled pool of human capital. New technologies bring in transparency in government and will also improve financial, commercial and industrial aspects; improve the shape of life for ordinary citizens – even if they are not in the digital surroundings.

As shown in Fig. 2, a majority of the investments will be made by the private entities or foreign investors which would require huge funding. Now since most of the developmental work comes under public infrastructure and services, banks may not be in a position to fund such projects. The organisation can grant more FSI to the private stakeholders for other projects in the same zone giving way to a smooth flow of funds.

The valuation of infrastructure, which grows manifold, needs to be shared by the society and the Government to support infrastructure development, ensuring a balanced growth. Today, the valuation due to infrastructure growth is not channellised towards infrastructure development and thus results in inequalities.

Conclusion

The concept of the Blue Economy is relatively new and marks a new era of development. The development agenda and the Sustainable Development Goals (SDGs) have been adopted by governments across the globe. The development of Blue Economy infrastructure projects with private capital through Public-Private Participation (PPP) route has become one of the common strategies in the developed and developing countries. A strategic planning for long-term transport and access infrastructure for a destination needs to take into consideration a destination's geographical location, community needs and values, planning and legal restrictions and investment potential as well as the potential ecological, economic and social impacts of development. Tourist destinations can utilise ICTs to strengthen their competitiveness and capture profitable market segments.

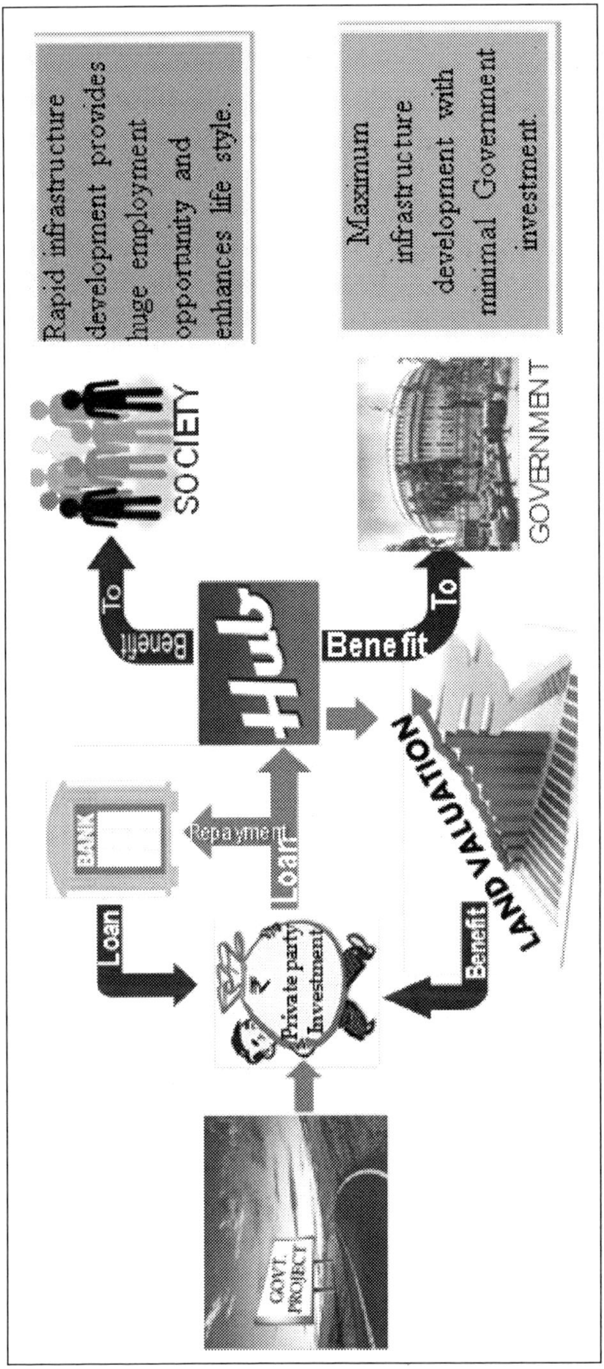

Figure 2. Secured Governance Funding Mechanism to develop HUBs or Smart Cities

The capital-intensive nature of basic infrastructure and competition for limited government budgetary resources have prompted governments to invite private investors to fulfil the widening demand–supply gap for infrastructure while the governments are endeavouring to meet the social commitments within the fiscal constraints. Fortunately, many property owners are able to capture the increase in land values due to public investment in the infrastructure development. A secured governance strategy implies that if the government uses this unearned increment of land value in a systematic way it would result in lower amounts of public expenditure in physical infrastructure, given that the increase in land values would provide the necessary financing mechanism.

REFERENCES AND NOTES

1. Handbook on Fisheries Statistics, 2014, Department of Animal Husbandry, Dairying & Fisheries, M/o Agriculture: Page No. 11. http://www.indiaenvironmentportal.org.in/files/file/Water percent20& percent20Related percent20Statistics percent202015.pdf
2. Ministry of Earth Sciences, http://dod.nic.in/programmes/geoscientific-studies-indian-exclusive-economic-zone
3. Cruise Shipping and Lighthouse Tourism: http://www.maritimeinvest.in/cruise-shipping-lighthouse-tourism
4. IBEF: http://www.ibef.org/industry/ports-india-shipping.aspx
5. Maritime India Summit 2016: http://maritimeinvest.in/ship-building-repair-recycling
6. http://www.thefishsite.com/fishnews/27617/indias-2016-marine-fish-production-below-estimations/
7. IBEF: http://www.india-opportunities.es/archivos/publicaciones/Oil-and-Gas-January-2016.pdf
8. EAI: http://www.eai.in/ref/ae/oce/oce.html
9. Ministry of New and Renewable Energy: http://mnre.gov.in/schemes/new-technologies/tidal-energy/
10. http://eprints.soton.ac.uk/349891/1/__soton.ac.uk_ude_PersonalFiles_Users_asv1a09_mydocuments_FSI percent20Admin percent20Files_LRET percent20Collegium_LRET percent202012_AAAA percent20The percent20LRET percent202012 percent20Books_LRET percent20Collegium percent202012 percent20Final percent20Version percent201-5_Book_vol_4.pdf
11. Press Information Bureau Government of India Cabinet: http://pib.nic.in/newsite/PrintRelease.aspx?relid=117691

12

Role of Entrepreneurs and Start-Ups in the Marine Industry: Enabling Through a 'Maritime Cluster' Approach

Suresh Bhardwaj

This chapter approaches the role of entrepreneurs and start-ups in the marine industry through a 'maritime cluster approach'. It dwells on the cluster theory and through its application analyses practical performances of some maritime clusters in the European countries. It identifies those specific enabling features of maritime cluster that create a fertile ground for entrepreneurship and start-ups to take root and grow faster than otherwise in the diverse domains of the 'Blue Economy'. Three main upgrading mechanisms are identified, which also become a source of self-reinforcing growth of clusters – innovation pressures, knowledge diffusion, and complementarity. They are with much of the dynamism of clusters deriving from 'economies of scope' rather than static 'economies of scale'.

Introduction

The most enabling feature of the Blue Economy is a 'maritime cluster' that can address adequately the 'role of entrepreneurs and start-ups'. There are some studies on the economic importance of such clusters, which get boosted by innovation, skill sharing and overall capacity development. This has been proven in many countries like Denmark, Norway and Netherlands where maritime clusters contribute immensely to the GDP of the country and support an enabling environment for entrepreneurs and start-ups that cut across the various segments of the industry, from manufacturing to software to services. This chapter analyses the role of a cluster as an enabler on two specific areas:

(a) What are the advantages for entrepreneurs and start-up firms to locate in a cluster, and
(b) What cluster-specific factors influence the development and provisioning of an enabling environment for entrepreneurs and start-ups.

The Cluster Theory

Clusters are defined as 'geographic concentrations of interconnected companies, specialized suppliers, service providers, firms in related industries, and associated institutions in particular fields that compete but also cooperate'.[1] While the economies resulting from agglomeration date back to the evolution of the neoclassical school of economics as a fundamental rationale for forming clusters,[2] it was not until the 1980s and it was later that its concept and impact on the economic performance was refined beyond cost-efficiency advantages. Krugman,[3] as an exponent of the new economic geography, and Porter, recognised the dynamic nature of clusters holding potential for sustained competitive advantage. Porter's diamond model as shown below in Fig. 1 conceptualises the dynamic elements contributing towards national competitiveness.

Clusters improve the 'factor conditions', which is the inputs that firms need in their operations, such as specialised labour and skills, targeted capital, infrastructure, technologies, research, education and knowledge. The greater concentration leads to more intense competition, simultaneous investments and innovation in the cluster industries, and the opportunities for cooperation, which will influence the firms' strategy, structure, and rivalry in the competitive environment and increase pressures to innovate the business processes. Companies are attracted to the cluster and new firms are established in order to serve the needs of this larger concentration of companies, which enables the growth of smaller businesses, whose costs are lowered by this proximity and the ability to outsource non-core activities. Thus, clustering enables the presence of advanced, specialised suppliers and others in related and supporting industries. Lastly, these favourable 'demand conditions' enable growth and facilitate innovation, as sophisticated customers pose intense requirements and needs, which increase the competitiveness of the suppliers and enable entrepreneurship and start-ups.

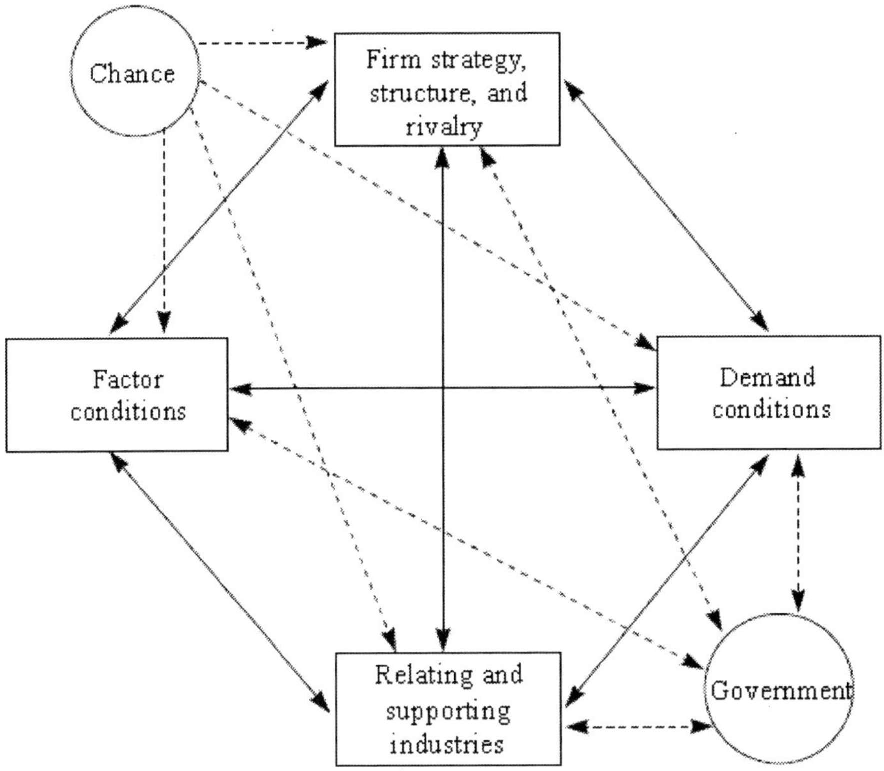

Figure 1. Porter's diamond model of cluster dynamics[4]

Leader firms facilitate the growth and beneficial dynamics within a cluster. Such companies have the size, market position, resources, and the entrepreneurial orientation that contributes positively to the competitiveness of the cluster and its population. Leader firm behaviour includes, among others, acting as lead user by posing sophisticated demands and high standards to suppliers, improving transfer of knowledge and the organisational infrastructure, encouraging internationalisation of partners and improving the internal labour market.

Performance Analysis of Some European Maritime Clusters

The Dutch Maritime Cluster

The Dutch Maritime Cluster has eleven maritime sectors constituents – Seaports, Maritime Services, Shipbuilding, Maritime Supply Industries, Water Sports, Fisheries, Dredging, Offshore, Navy, Inland Shipping and Seagoing Shipping. The funding is by member contributions and its main function is to lobby for its members at various levels of the government – local, regional, national and European level – either directly or as members of the European and global associations. The 11 different yet complementary industries comprise 11,850 companies that operate within a 100-kilometre radius. This encourages close cooperation on innovation and production, allowing the industries to build on each other's strengths. The government, academia and private sector cooperate to implement long-term innovation programmes. Innovation is a prime responsibility of the entrepreneurs, but the government has created a number of generic instruments to stimulate innovative behaviour and the innovative capacity of the people and companies. From the life cycle perspective, this cluster had the ability to re-invent itself and ignite newer phases of cluster dynamism in the wake of changing circumstances in the shipping industry.[5]

The Norwegian Maritime Cluster

Norway has 10 per cent of the world merchant fleet, placing it in the top three world ranking, and carries out 15 per cent of the global oil exploration activities on the near-shore. The main components of the Norwegian cluster are: maritime shipping, marine equipment suppliers (mainly for the offshore oil and natural gas); maritime services (finance, insurance, brokering, maritime law, classification and certification of ships, port services); shipbuilding (specialised vessels for oil prospecting and exploration, highly sophisticated cruise ships, factory ships and fishing vessels, including equipment for propulsion and navigation, patrol boats, specialised vessels for the transportation of chemicals and liquefied natural gas, icebreaker vessels) and fisheries. All these sectors, especially those related to shipbuilding and equipment/marine machinery, are characterised by a strong R&D intensity, involving companies, universities and public R&D centres. Norwegian companies cooperate with other firms around R&D and the supplier–buyer cooperation to improve innovation is particularly strong. The relationships between the industries in the same sector are strong and there are strong relationships between what is traditionally labelled as shipping (shipping companies, ship brokerage services, insurance, bank/finance, classification, consultants, other services) and the shipping supply industry (shipbuilding, equipment, engines, wholesale).

Danish Maritime Cluster

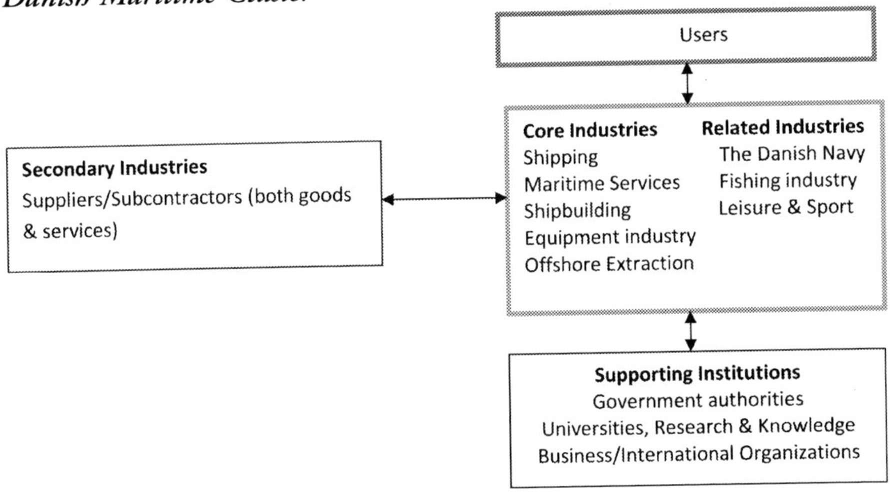

Figure 2. The Danish Maritime Cluster[6]

The strongest interdependency in the Danish Maritime Cluster as shown in Fig. 2 is the socio-economic element, and education and a common labour market in particular. The fact that maritime officers return to land after serving a certain period on ships, and take up employment in the shipping companies' land-based organisations and in the other maritime industries, is seen as an important driver for growth. A number of maritime parties highlight the Danish Maritime Authority as an important factor for increasing interaction and dynamics in the maritime cluster. Innovation is also seen as an important driver for greater interaction. For many maritime enterprises, user-driven innovation is the dominant element. Here, it is the dialogue between suppliers and the customer that contributes to the development of processes and products. In other areas, emphasis is on research-driven innovation – where research is the focal point.

In 2005, the Danish Government established the Danish Maritime Trust Fund, which is intended to provide financial support for initiatives and measures that develop the maritime cluster – Blue Denmark. Research, development and innovation are key focus areas for the Trust Fund. The Trust Fund owns a number of shares in the Danish Ship Finance, a private shipping finance company, and the dividends from these finance the Foundation's grants. The aim of the work in cluster organisations is to focus on and develop dynamics in the cluster. The competitiveness of the individual industries depends on the other industries, along with links to universities, maritime colleges/universities and other educational institutions, government authorities and business organisations.

Economic Significance of the Maritime Clusters

It is thus evident that clusters consist of related economic actors and institutions active in multiple levels of cluster, and that the scale is sufficient to achieve the critical mass, which leads to the development of specialised suppliers, resources and services.[7] At the heart of the maritime cluster has been the pursuit of an enhanced awareness of the economic contribution of maritime-related sectors to a national or regional economy.

Four agglomeration economies are widely recognised of the cluster: a joint labour pool, a broad supplier and customer base, knowledge spillover, and low transaction costs. The total economic impact of maritime activities may be broken down into *direct* and *indirect* effects. *Direct* considers the economic activity and jobs generated directly by the cluster. *Indirect* effects include the jobs and demand created up and down the supply chain, as well as the consumption generated in the rest of the economy. Thus, the value and demand generated in the clusters trickle down through the overall economy to create further jobs and demand, making the combined economic importance of an even greater nature than simply direct activities. This explains why firms increasingly locate in the cluster.[8]

According to the European Cluster Observatory data 2016, economic activities located clusters to account for 39 per cent of the jobs and 55 per cent of the wages. Significantly, they also account for 87 per cent of patents, proving them to be innovation and growth drivers.

Enabling Environment for Entrepreneurship and Start-Ups

Clusters play a significant role in innovation. The modern approach to innovation is less linear, emphasising the benefits of 'open innovation' in dynamic, networked environments, where organisations and institutions in different levels of a cluster interact to combine knowledge to new products and processes. Clusters help to close the gap between business, research and resources, thus facilitating the assimilation of knowledge. Proximity plays a vital role in innovation, as it is at the level of the region that employees, businesses and organisations interact with one another, particularly so the innovative small to medium-sized enterprises.

Entrepreneurship and start-ups are a major driver of innovation, competitiveness and growth as they can have better presence in the key sectors such as service and knowledge-based activities. Small businesses depend on entrepreneurs – the individuals who have the ideas and are willing to take risks necessary to get firms off the ground. Ideas need to be transferred into real innovation – a new economic activity that creates job and wealth. A cluster environment enables promoting business incubators and growth of small innovative businesses with enabling cohesion policies among the actors that offsets risks. Clusters serve as an important tool to improve the business environment particularly for the SMEs, entrepreneurs and start-ups as it facilitates a change towards regional competitiveness by upgrading traditional economic activities through innovation and facilitating the emergence of new products, processes and services which are competitive in the global market. It builds on local strengths and helps create synergies between businesses and respond to market opportunities. Clusters help to 'glue' multinational companies and SMEs and root them to the region through favourable local collective competition goods created in the cluster environment.[9]

Entrepreneurship is thus necessary to ensure that innovative ideas can be turned into new products and services that create growth, quality jobs and help address societal challenges. It fosters a smart, sustainable and inclusive growth and must be flagged as a strategic initiative. A useful concept in this regard is the 'triple helix', which holds that innovation networks in clusters depends upon academic and research institutions (universities), companies, capital and entrepreneurship (private sector) as well as favourable framework conditions (public administration).

EU Start-Up and Scale-Up Initiative

The European Union commits to provide entrepreneurs with favourable business environment through its cluster policies and smart specialisation strategies as they help to take the geographic and thematic context into account in order to boost jobs, SME growth and investment. The EU believes that the implementation of these strategies promises to trigger an industrial modernisation process and create more productive, diversified and resilient economies. It should in particular contribute to providing SMEs – which are the backbone of economy - with better support and investment opportunities.

The EU has specifically focused on the development of entrepreneurship and start-ups by a 'Start-Up and Scale-Up' initiative.[10] It seeks to achieve this by:

(a) Helping them navigate the – often perceived – regulatory barriers to fully benefiting from the Single Market,
(b) By making it easier for them to have a second chance, without being stigmatised if their idea doesn't succeed the first time around, and
(c) By improving access to funding by boosting private venture capital investment.

The initiative brings together a range of existing and new actions to create a more coherent framework to allow the start-ups to grow. The initiative also puts emphasis on helping ease up regulatory requirements, improving innovation support and fostering ecosystems where start-ups can connect with potential partners such as investors, business partners, universities and research centres.

Heterogeneity in Clusters and 'Economies of Scope'

It is important for clusters to consist of a critical mass of heterogeneous activities, in order to make the value of the cluster bigger than the sum of its parts, as well as enable an increased specialisation of companies. Critical mass matters; the more companies there are in a certain set of industries the more likely it is that higher levels of productivity and innovation will be achieved. This is partly due to specialisation. If there is a large local market, companies can focus more on being very good at a specific narrow activity, and employees can focus more on acquiring very specialised skills. It is also partly due to competitive pressure. If there is more competition among peers for customers, employees and ideas, each of them has to work harder to develop unique products and services that stand out in the market. This puts pressure on companies, which can be challenging, but it makes them more likely to succeed in the national and global competition.

By 'heterogeneous', it is meant that cluster actors should be well distributed across vertical and horizontal dimensions, so that the different levels of the supply chain are represented, along with a number of activities horizontally within each sector. While clusters reflect the specialisation of locations in specific economic activities, they capture two important additional aspects. First, clusters reflect specialisation in groups of related industries, not just one narrow activity. Much of the dynamism of clusters derives from these 'economies of scope' rather than static 'economies of scale'. Clusters are thus much more a reflection of cross-industry linkages and 'related diversification' than of narrow specialisation. Second, clusters reflect the dynamic interaction between a large number of firms. Specialisation through the presence of one large firm or plant does not constitute a cluster.

Heterogeneity and Life Cycle of a Cluster

A growing literature looks at the life cycle of clusters. Clusters often seem to follow an S-shaped development path. After an (often long) phase of slow gestation, a cluster reaches a size where cluster effects set in and growth accelerates. This growth than becomes self-reinforcing; cluster effects reach their full scale and growth explodes. Eventually, the growth moderates as the cluster reaches its market potential and congestion effects become more relevant. Some clusters then manage to reinvent themselves, finding a new market or technology to ignite a next phase of cluster dynamisms. Others, however, get locked into existing technology and eventually shrink, as their markets disappear or other locations develop more dynamism. Holte and Moen describe how the positive factors (increasing embeddedness and inter-relatedness generate increasing returns and stimulate rising economic performance) turn into a negative lock-in with an inflexibility and reduced innovation ability, as illustrated in Fig. 3.

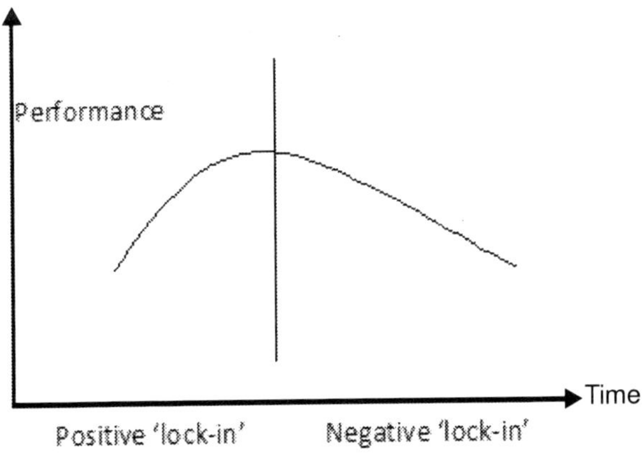

Figure 3. Illustrating development paths in the life cycle of a cluster[11]

According to Menzel and Fornahl the connections between the quantitative and qualitative development of the cluster indicate that its heterogeneity of knowledge is the foundation of its development.[12] The cluster declines if its heterogeneity cannot be sustained. If the heterogeneity increases again, the cluster moves back in the cycle and enters a new growth stage. This increase in heterogeneity can be incremental, for e.g., the integration of new knowledge from the respective technological trajectory into the cluster.

The Formation of Clusters and 'Cluster Organisations'

Clusters are in the vast majority of cases not 'created': they emerge, because different locations provide different types of opportunities for specific companies to invest, succeed and grow. Clusters are the result of a cumulative process, where the success of one company paves the way for others to follow. Such processes take a long time, and are inherently unpredictable. Cluster evolution is a natural process, but it is not automatically a successful one. In most cases, success depends on creating specific qualities of a business environment that give a location a unique and lasting advantage. By helping to create these qualities, governments can have a significant influence on the emergence and growth of a cluster. The challenge is to invest in creating the kind of qualities that meet a market need. Often, the difference has been that successful regions have been able to integrate the investment into a broader set of qualities that made their location attractive for a specific cluster.

'Cluster organisations' are organisational platforms set up to facilitate collective action within the cluster and provide a way to tap into the tools that the government provides at different levels for the cluster. Different organisational structures have emerged to play this role; some countries have created quasi-private cluster organisations in the public sector that provide the cluster with 'one-stop' access to all relevant government programmes, and are reliant for their funding on their ability to mobilise private sector activities that provide co-funding.

There are basically two possible approaches for maritime cluster organisations as to the initiative to establish a cluster. This initiative can be categorised in government-induced organisations (top-down) and cluster organisations induced by leader firms and/or sector associations (bottom-up). The main strength of top-down maritime cluster organisations is that they are initiated by the government and consequently aim to increase the growth of all maritime sectors within the cluster. This growth is supported by a long term strategy and policy, focusing on the improvement of research, development and innovation and the creation of a level playing field through legislation within the maritime cluster. Through R&D intensity and legislation and relatively high budgets, top-down clusters can positively influence common interests.

The main strength of bottom-up cluster organisations is the enhancement of business efficiency and opportunities. This strength goes hand in hand with a strong focus on finding solutions for operational problems. Moreover, cluster organisations are structurally evaluated by their members with implications on membership contributions. Also there is the ability of the cluster to operate as a single-point-of-entry to the government initiatives. Bottom-up cluster organisations often take the lead in organising events (e.g. to promote working in the cluster) and in establishing issues platforms (e.g. to exchange best practices and lessons learned).

A weakness however, of bottom-up clusters concerns the differences in sector interests, sometimes leading to a difficult decision-making progress within the cluster organisations. This difference in sector interests often results in a rather narrow focus on softer themes (e.g. promotion). Furthermore, company and sector interests of bottom-up clusters are often not balanced. This potential imbalance is sometimes caused by differences in the weight and influence of members or supporting companies and sectors within the cluster organisations. In particular, this imbalance is often caused by the strong impact of the largest contributing members in the decision-making process. Another weakness is that bottom-up cluster organisations are sometimes not always much focused on strategy and the long-term growth of the sectors involved and on the cluster as a whole.

The main threat for cluster organisations is that the too narrow local focus may lead to the suboptimal functioning of a maritime cluster organisation. Also, the limited attention for the role of (relatively) smaller sectors and companies within the maritime cluster poses a threat to the bottom-up cluster organisations. In terms of funding, there are differences in the organisation and/or associations that financially contribute to the cluster organisations. Also, significant differences exist between the level of the budgets that top-down and bottom-up cluster organisations receive.

Top-down maritime cluster organisations exist in Germany and Denmark. These two organisations fund their activities with government budgets. The budgets of top-down maritime cluster organisations are far higher than those of bottom-up maritime cluster organisations, because they include the overall budgets of the government administration and R&D and/or maritime education programmes as well.

The main goal in establishing cluster organisations is the installation of a structure that provides clear solutions for cluster-issues (e.g. labour market, innovation, developing entrepreneurship and start-ups), regardless of the initiator of that cluster organisation. The main issues at the basis of the establishment of a maritime cluster organisation are to increase competitiveness, to promote maritime sectors and to improve coordination within the cluster. Competitiveness on the cluster level can be measured by looking at the market shares, the growth of value added and the gross production of a cluster. The organisational structure of these organisations depends on a different climate and/or environment influenced by traditions, history, culture, governmental structure and driving sectors.[13]

Conclusion

The main contribution of Porter's theory has been the outcome of four interlinked advanced factors (the factor conditions, the demand conditions, the related and supportive industries and the government) in and between companies within clusters, which can be influenced in a proactive way by government. To substantiate on the role of 'entrepreneurship and start-ups' in the maritime industry, a differentiation framework for maritime clusters was chosen as a model, because it allows the evaluation of how these factors interact with each other to form conditions where innovation, competitiveness and other enabling conditions occur to stimulate their role.

Three main upgrading mechanisms are the sources of a self-reinforcing growth of clusters: innovation pressures, knowledge diffusion and complementarity. Competition, imitation and sophisticated demands from customers lead to pressures to innovate, which spread through the supply chain. Knowledge is diffused through the levels of a cluster through a high mobility of human resources and forums for communication and interaction, as well as cooperation. Finally, growth and establishments of new companies, mainly start-ups and SMEs, lead to a critical mass, which enhances synergies and the profitability of projects, and makes the cluster an even more attractive location. In addition, clusters provide a fertile combination of entrepreneurial dynamism and contribute to building a knowledge-based economy.

The concept of clusters goes beyond geographical proximity in that it emphasises the relationships and social ties among the actors in the cluster population. The personal interaction, frequent communication and a sense of a common identity that can arise out of a cluster's diverse networks, from project cooperation to cluster organisations and networking arrangements, are what make up the linkages that foster the cluster dynamics. While some relationships are contractual, many are founded upon a built-up trust between the cluster actors, over a long time. Such relationships and the proximity can arguably increase both intentional and unintentional resource sharing, in terms of formal and informal knowledge spillovers. These positive externalities from the co-location tend to attract more economic actors to a cluster.

Overall, institutional settings play a significant role in enabling competitiveness and expansion, retaining maritime activities and attracting companies, as well as increasing the level of innovation. This includes government policy, interest organisations, educational institutions and research bodies. A cluster-based policy shifts the role of the government towards building appropriate legislative frameworks for cooperation and growth.

Successful clusters reinforce and upgrade themselves, driven by the dynamic inter-relationships of cluster actors in terms of competition, cooperation, knowledge spillover and innovation. The benefits of such concentration of activities are largely undisputed, as regions worldwide compete to provide the best framework conditions to attract and retain businesses, investments and human resources.

REFERENCES AND NOTES

1. Porter, Michael E., 'Clusters and the New Economics of Competition', *Harvard Business Review*,1998, November-December.
2. Marshall, A., *Principles of Economics*, 8th edition, Basingstoke: Macmillan, 1890.
3. Krugman, P., *Geography and Trade*, Leuven: Leuvan University Press, 1991.
4. Porter, Michael E., *The Competitive Advantage of Nations*, London: Macmillan, 1990.
5. Monterio, P., Noronha T. and Neto P., 'A Differentiation Framework for Maritime Clusters: Comparisons across Europe', *Sustainability*, 5, 2013, 4076-4105; doi:10.3390/su5094076
6. Ministry of Economic and Business Affairs, *The Danish Maritime Cluster-An Agenda for Growth*, Ministry of Economic and Business Affairs, 2006, Copenhagen.
7. European Commission, 'Towards World-Class Clusters in the European Union: Implementing the Broad-Based Innovation Strategy'.
8. 'The Economic Significance of Maritime Clusters: 2006', Danish Shipowners Association.
9. OECD, *Competitive Regional Clusters*, OECD Reviews of Regional Innovation, 2007.

10. EU, *Smart Guide to Cluster Policy*, 2016, doi:10.2873/48105

11. Holte, E. and Ø. Moen, 'Successful Maritime Clusters: Key Drivers and Criteria', IGLO-MP 2020 Working Paper 01–2010, Trondheim; Norwegian University of Science and Technology, 2010.

12. Menzel, M.-P. and D. Fornahl, 'Cluster Life Cycles – Dimensions and Rationales of Cluster Evolution', *Industrial and Corporate Change*, 19, 2010, pp. 205-238.

13. Viederytea, R., 'Maritime Cluster Organizations: Enhancing Role of Maritime Industry Development', *Social and Behavioural Sciences*, 81, 2013, pp, 624-631.

13

The Blue Economy: Summary and Takeaways

Vasudha Chawla and Kapil Narula

The conference theme aptly reflects the growing interest in the Blue Economy. The conference was divided into five sessions and based on the views of the papers presenters and discussions, the session-wise summary and key takeaways from the event are as follows.

Blue Economy and Sustainable Development

The first session set the backdrop for the conference and various interpretations of the Blue Economy were discussed. This resulted in developing a common understanding of the concept and also helped in highlighting the contemporary relevance of the Blue Economy. The underlying principles, drivers and the means to implement the Blue Economy in the context of achieving sustainable development and balanced growth were discussed. It was emphasised that oceans need to be seen as 'development spaces' and as 'common heritage of mankind' where socio-economic development is decoupled from environmental degradation. It emerged that the Blue Economy promotes zero waste, redistribution and social inclusion and hence contributes to different Sustainable Development Goals (SDGs) directly and indirectly. Further, the Blue Economy and SDGs are aligned and mutually reinforcing concepts and efforts are underway to translate the SDG 14 which is focused on oceans, from 'ambition to action', in the forthcoming UN High Level Conference to be held at New York in June 2017.

While different countries are attempting to develop the Blue Economy, they are at the initial stages of implementation of the Blue Economy. It emerged that the concept of the Ocean Economy holds much more relevance for the Small Island Developing States (SIDS) than the Blue Economy. The SIDS can develop Ocean Economy through measures like port development, seafood processing and promoting marine tourism. It was highlighted that a common authority for coordination is important and can play an important role in the development of the Blue Economy. For instance, Mauritius has established a Ministry of Ocean Economy in 2014, and a National Ocean Council was constituted to formulate the National Ocean Policy. However, the country faces challenges such as an absence of a proper legal framework, lack of data, etc. Nevertheless, the SAMOA (Small Island Developing States Accelerated Modalities of Action) Pathway and the Mauritius Strategy for the Further Implementation of the Programme of Action for the Sustainable Development of SIDS (Small Island Developing States) has paved the way for a sustainable development of SIDS.

There are a number of institutions and mechanisms for international cooperation on the Blue Economy and the 1982 United Nations Convention on the Law of the Sea (UNCLOS) sets the basic international legal framework for the Blue Economy. Various other treaties and mechanisms, and regional institutions such as the APEC, EU and ASEAN, are contributing towards the development of the Blue Economy; however, its implementation remains a challenge. The promotion of the Blue Economy in a top-down manner is also beset with the absence of a single international institution and is hindered by fragmented facets of the existing mechanisms. Thus, there is a need to develop a 'common institution' to promote and regulate the Blue Economy initiatives and these institutions can implement mechanisms for undertaking a shared responsibility for an effective management of the seabed resources. The Asian Infrastructure Investment Bank (AIIB) and the Belt and Road Initiative (BRI) are a few initiatives which have the potential to promote the Blue Economy.

Habitat Protection and Ocean Ecology

Oceans are an important source of animal and plant protein but the 'high seas' are perhaps the least protected areas on the planet. The indiscriminate fishing and destruction of marine biological resources has gradually lowered the capacity of the oceans to nurture marine life, leading to reduced ocean productivity. Habitat protection and marine ecology are therefore critical factors for the development of Blue Economy. Marine Protected Areas (MPAs), Integrated Coastal Zone Management (ICZM) and Marine Spatial Planning (MSP), which are essential for enhancing the Blue Economy.

Fisheries and aquaculture play a major role in protecting the ocean ecology. Fisheries are expected to increase by 300 per cent and the fish cultivation is expected to increase to nearly 65 million tonnes through aquaculture. However, a coherent and consistent regulatory framework and a strong national infrastructure are essential for realising new opportunities in the fishing sector.

Coastal and marine tourism are important components of the Blue Economy and Seychelles has demonstrated leadership in propelling a new vision for the Blue Economy. The island nation has been successful in protecting and restoring coral reefs from bleaching and has launched foundational projects for the Blue Economy – 'The Marine Spatial Plan' and adapting MPAs to climate change. It has also adopted innovative financing mechanisms for funding the marine conservation projects.

It was emphasised that the Indian Ocean is prone to the overexploitation of resources, threats from pollution, impact of climate change and maritime security disputes. There is also a need to better manage the resources in the region. The importance of MSP for protecting marine ecosystems and the ecosystem-based management (EBM) approach for habitat protection was also highlighted.

Ocean Governance and Marine Resources

There are both legal and governance gaps in the Exclusive Economic Zone (EEZ) and in the Areas Beyond National Jurisdiction (ABNJ). The 'High Seas' are the 'common heritage of mankind' and even landlocked countries have equal rights and responsibilities in these areas. Offshore hydrocarbons, gas hydrates and natural gas, minerals from deep seabed mining, marine renewable energy from waves, tidal and offshore wind, and marine bioprospecting offer immense potential. Although extraction of these resources is important for the benefit of humanity, there are associated environmental impacts.

Seabed mining has emerged from the scarcity of mineral resources on land. Seabed mining dates to as far back as in 1965, where the viability of extraction of copper, nickel and cobalt from nodules found on the sea floor was established. However, one of the major challenges of seabed mining at that time was the lack of expertise to mine, process and extract the produced metal. Over time, technology provided solutions to these limitations, but it is important that in anticipating the quantum of seabed minerals like manganese nodules and sea floor sulphides, the economic viability of commercial extraction is established. The International Seabed Authority (ISA) regulates the exploration of deep seabed mining. India recognises the opportunities to extract mineral resources in the Indian Ocean and is well aware of the growing resource competition in seabed mining. The sector is growing, and there are 26 contracts that have been signed so far: 16 for polymetallic nodules; six for sulphides and four for manganese. India has signed a contract for 15-year exploration of polymetallic sulphides in the allotted area which formalises India's exclusive rights for exploration in the Central Indian Ridge and South West Indian Ridge in the Indian Ocean. The competition for deep seabed mining is growing and it is India's endeavour to engage in mutual cooperation. Some of the challenges for seabed exploration and commercial production are transfer of technology, financial mechanisms and regulations which have led to a slow pace of developments in the sector.

Ocean energy is an emerging sector and can contribute to the Blue Economy by providing clean and renewable energy. Ocean energy including tidal waves and thermal energy are a promising source of clean energy and can support growth. However, a lack of awareness about the potential of ocean energy, the perceived high cost of energy and the fear of environmental issues poses barriers towards the adoption of ocean energy. Most of the equipments that have been developed are for temperate climate and thus need to be tuned to the requirement of tropical climate, before deploying the technology in the region. The promotion of domestic firms and the development of local technology will help in streamlining the supply chains and will bring down the cost of associated machinery, thereby lowering the cost of generated electricity.

Ambiguity between the 1982 UNCLOS and the International Court of Justice (ICJ) remains a major challenge in the settlement of maritime disputes. Until 1960, disputes related to the maritime domain were settled by the ICJ, but proliferation of a number of institutions for the settlement of maritime disputes led to new challenges as these institutions can only work if both parties want to agree to accept and abide by the ruling of the courts. It was espoused that the legal international system has been tweaked and has been observed to work in the interest of major powers. Due to the lack of one unilateral institution, the present legal system is inadequate to deal with maritime disputes.

Maritime Security, Industry and Administration

Maritime security is vital for the development of the Blue Economy. The security of maritime infrastructure such as shipping, offshore oil platforms and ports, is essential, given that these contribute to the goals of the Blue Economy.

Maritime security faces both traditional and non-traditional challenges and maritime boundary disputes are a hindrance to the realisation of the shared responsibility. Although, there are many stakeholders in maritime security, the lack of a proper framework is a challenge to coordinate between these disparate stakeholders. India's cooperation with other countries such as Dosti (Maldives and India), Tri Dosti (Maldives, India and Sri Lanka), and the recent MoU signed between India and Bangladesh are good examples of India's cooperative initiatives. The signing of Standard Operating Procedures (SOPs) can further lead to on-ground cooperation between the countries and law enforcement and Search And Rescue (SAR) operations can be other areas of cooperation.

The marine leisure industry is an enabler of the Blue Economy as it directly and indirectly results in the creation of jobs. It also benefits entrepreneurs and promotes the local industry. Despite a long coastline and waterways, the boating industry in India is not a thriving industry. It faces challenges from the lack of infrastructure as also incentives. Public-Private Partnerships (PPP) models and single-window clearance will help to grow the boating industry from the bottom up.

Streamlining the marine administration is also essential for the Blue Economy. A national level platform is required for administering and coordinating maritime issues across various sectors and the European model of marine administration can be explored. It was proposed that a leader with sound knowledge on maritime issues may be positioned as the Chief Maritime Advisor to coordinate and oversee the implementation of national policies.

Enablers of the Blue Economy

Apart from various thematic issues and sectors of the Blue Economy, there are common issues such as policies, funding, business models, innovation and technological developments which act as catalysts for enabling the Blue Economy.

The economic potential of the blue economy lies in smart port cities, marine tourism, marine transportation, ports, the shipbuilding industry, fisheries, offshore oil and gas, ocean renewable energy, marine bio-industry and deep sea mining. The concept of 'secured governance' was proposed to capture the hidden value of the Blue Economy. The strategy proposes that if the government utilises the locked-in economic value of land in a systematic manner, it could result in a lower amount of public expenditure for building infrastructure. The increased land value would then be able to provide the necessary financing for undertaking projects and this funding mechanism to develop smart port cities, can augment the Blue Economy.

Technology's role as an enabler of Blue Economy was emphasised. In India, there are several opportunities, especially for harnessing the mineral resources, renewable and non-renewable energy resources, food, fresh water, etc. However, there are many technological challenges in the underwater environment. These are: high atmospheric pressure, temperature, hydrodynamics, lack of natural light and poor communication and these demand the appropriate technologies. Weather prediction and monitoring of the physical parameters underwater also play an important part in the success of these technologies. The National Institute of Ocean Technology (NIOT) has been successful in creating technologies such as the deep sea in-situ soil tester, Remotely Operable Vehicle (ROV)-Sagar Nidhi, Rodrigues Triple Junction for live underwater streaming, and developers are currently working on a manned submersible vehicle for underwater exploration.

Entrepreneurs and start-ups play a critical role in harnessing the Blue Economy, and this can be achieved through a Maritime Cluster Approach. European models like the Dutch Maritime Cluster, the Norwegian Maritime Cluster and the Danish Maritime Cluster are successful examples of application of such models. The use of Porter's theory and the outcome of four interlinked advanced factors – the factor conditions, the demand conditions, the related and supportive industries and the government were proposed to develop the maritime clusters. Enabling mechanisms like innovation and competitiveness, and heterogeneous activities like the economy of scope and the lifecycle of the cluster are important factors which need to be considered. There are two possible cluster organisations – namely, government-induced organisations (top-down) and leader firms (bottom-up) which were proposed for promoting entrepreneurship and start-ups in the maritime industry in the Indian context.

Conference Takeaways

(a) **Session 1: Blue Economy and Sustainable Development**

- (i) The Blue Economy and Sustainable Development Goals are aligned and mutually reinforcing.
- (ii) The 1982 UN Convention on the Law of the Sea (UNCLOS) sets out the basic international legal framework for the Blue Economy.
- (iii) International institutions, treaties and mechanisms to develop the Blue Economy are in place, but the actual implementation of mechanisms is challenging.
- (iv) Mauritius is diversifying towards port development, seafood development and tourism but endorses the concept of Ocean Economy rather than Blue Economy.

(b) **Session 2: Habitat Protection and Ocean Ecology**

- (i) There is a threat to fisheries around the world and a coherent and consistent regulatory framework with a strong national infrastructure is required to support sustainable fishing, both in coastal waters and in the deep sea.
- (ii) Seychelles' economy is driven by coastal and marine tourism and it has been the leader in propelling the Blue Economy. Various projects have been implemented to promote the Blue Economy such as innovative funding mechanisms and the restoration of coral reefs around the island nation.
- (iii) Marine Spatial Planning is necessary in the Indian Ocean Region. Collaboration amongst countries and an Ecosystem Based Management (EBMs) approach is suggested for furthering the implementation of the Blue Economy.

(c) **Session 3: Ocean Governance and Marine Resources**

- (i) The International Seabed Authority plays a major role in regulating Deep Sea Mining.
- (ii) While India has signed a contract with the ISA for deep sea

exploration in the central Indian Ocean region, there are various hurdles which need to be overcome prior to commencing commercial deep seabed mining in the area.

(iii) The marine renewable energy including tidal waves and ocean thermal energy has a large potential and can be harnessed in tropical waters to support the regional energy needs.

(iv) There is an ambiguity between international legal institutions and lack of one unilateral institution and a comprehensive legal framework remains a major challenge for the settlement of maritime disputes.

(d) Session 4: Maritime Security, Industry and Administration

(i) The maritime security faces traditional and non- traditional challenges which are a hindrance for the development of the Blue Economy.

(ii) The lack of a proper framework for the development of the Blue Economy inhibits cooperation and coordination between the stakeholders.

(iii) There is a need for an apex body for administering and coordinating maritime issues in India.

(iv) The marine leisure and marine tourism industry in India is at a nascent stage and needs to be promoted by the Indian government as its aids the Blue Economy.

(v) Public Private Partnership models and a creation of a single authority can help overcome the challenges of marine leisure industry in India.

(e) Session 5: Enablers of Blue Economy

(i) Secured governance can play an important role in achieving the objectives of the Blue Economy.

(ii) There are many opportunities in India to harness the Blue Economy.

(iii) India is working on various technologies for underwater data gathering and to augment underwater exploration.

(iv) A 'cluster approach', which includes entrepreneurs and start-up firms, helps in harnessing the potential of the Blue Economy and is recommended for India.

Conclusion

The conference achieved its goal to discuss the various facets of the Blue Economy and facilitated a common understanding of the concept. The various constituents of Blue Economy were discussed and the opportunities and challenges to implement the Blue Economy in the Indian context were identified. It was concluded that it is important to bring together various stakeholders and to integrate the efforts for harnessing the Blue Economy in a synergistic and holistic manner.

Appendix A

List of Sustainable Development Goals

Goal 1	End poverty in all its forms everywhere
Goal 2	End hunger, achieve food security and improved nutrition and promote sustainable agriculture
Goal 3	Ensure healthy lives and promote well-being for all at all ages
Goal 4	Ensure inclusive and equitable quality education and promote lifelong learning opportunities for all
Goal 5	Achieve gender equality and empower all women and girls
Goal 6	Ensure availability and sustainable management of water and sanitation for all
Goal 7	Ensure access to affordable, reliable, sustainable and modern energy for all
Goal 8	Promote sustained, inclusive and sustainable economic growth, full and productive employment and decent work for all
Goal 9	Build resilient infrastructure, promote inclusive and sustainable industrialization and foster innovation
Goal 10	Reduce inequality within and among countries
Goal 11	Make cities and human settlements inclusive, safe, resilient and sustainable
Goal 12	Ensure sustainable consumption and production patterns
Goal 13	Take urgent action to combat climate change and its impacts*
Goal 14	Conserve and sustainably use the oceans, seas and marine resources for sustainable development
Goal 15	Protect, restore and promote sustainable use of terrestrial

	ecosystems, sustainably manage forests, combat desertification, and halt and reverse land degradation and halt biodiversity loss
Goal 16	Promote peaceful and inclusive societies for sustainable development, provide access to justice for all and build effective, accountable and inclusive institutions at all levels
Goal 17	Strengthen the means of implementation and revitalize the global partnership for sustainable development

Appendix B

Sustainable Development Goal 14 (SDG14): Goals, Targets and Specific Indicators

Goal 14. Conserve and sustainably use the oceans, seas and marine resources for sustainable development.

Target 14.1 By 2025, prevent and significantly reduce marine pollution of all kinds, in particular from land-based activities, including marine debris and nutrient pollution.

Indicator 14.1.1: Index of coastal eutrophication and floating plastic debris density.

Target 14.2 By 2020, sustainably manage and protect marine and coastal ecosystems to avoid significant adverse impacts, including by strengthening their resilience, and take action for their restoration in order to achieve healthy and productive oceans.

Indicator 14.2.1: Proportion of national exclusive economic zones managed using ecosystem-based approaches.

Target 14.3 Minimize and address the impacts of ocean acidification, including through enhanced scientific cooperation at all levels.

Indicator 14.3.1 Average marine acidity (pH) measured at agreed suite of representative sampling stations.

Target 14.4 By 2020, effectively regulate harvesting and end overfishing, illegal, unreported and unregulated fishing and destructive fishing practices and implement science-based management plans, in order to restore fish stocks in the shortest time feasible, at least to levels that can produce maximum sustainable yield as determined by their biological characteristics.

Indicator 14.4.1 Proportion of fish stocks within biologically sustainable levels.

Target 14.5 By 2020, conserve at least 10 per cent of coastal and marine areas, consistent with national and international law and based on the best available scientific information.

Indicator 14.5.1 Coverage of protected areas in relation to marine areas.

Target 14.6 By 2020, prohibit certain forms of fisheries subsidies which contribute to overcapacity and overfishing, eliminate subsidies that contribute to illegal, unreported and unregulated fishing and refrain from introducing new such subsidies, recognizing that appropriate and effective special and differential treatment for developing and least developed countries should be an integral part of the World Trade Organization fisheries subsidies negotiation (taking into account ongoing World Trade Organization negotiations, the Doha Development Agenda and the Hong Kong ministerial mandate.)

Indicator 14.6.1 Progress by countries in the degree of implementation of international instruments aiming to combat illegal, unreported and unregulated fishing.

Target 14.7 By 2030, increase the economic benefits to the Small Island Developing States and least developed countries from the sustainable use of marine resources, including through sustainable management of fisheries, aquaculture and tourism.

Indicator 14.7.1 Sustainable fisheries as a percentage of GDP in Small Island Developing States, least developed countries and all countries.

Target 14.a Increase scientific knowledge, develop research capacity and transfer marine technology, taking into account the Intergovernmental Oceanographic Commission Criteria and Guidelines on the Transfer of Marine Technology, in order to improve ocean health and to enhance the contribution of marine biodiversity to the development of developing countries, in particular Small Island Developing States and least developed countries.

Indicator 14.a.1 Proportion of total research budget allocated to research in the field of marine technology.

Target 14.b Provide access for small-scale artisanal fishers to marine resources and markets.

Indicator 14.b.1 Progress by countries in the degree of application of a legal/ regulatory/policy/institutional framework which recognizes and protects access rights for small-scale fisheries.

Target 14.c Enhance the conservation and sustainable use of oceans and their resources by implementing international law as reflected in the United Nations Convention on the Law of the Sea, which provides the legal framework for the conservation and sustainable use of oceans and their resources, as recalled in paragraph 158 of "The Future We Want".

Indicator 14.c.1 Number of countries making progress in ratifying, accepting and implementing through legal, policy and institutional frameworks, ocean-related instruments that implement international law, as reflected in the United Nations Convention on the Law of the Sea, for the conservation and sustainable use of the oceans and their resources.

Contributors

Rear Admiral (Retd) Md. Khurshed Alam, is Secretary, Maritime Affairs Unit at Ministry of Foreign Affairs, Bangladesh. He performed the duties of the Deputy Agent and head of the legal team in the "dispute concerning delimitation of the maritime boundary between Bangladesh and Myanmar" and the "Bay of Bengal Maritime Boundary Arbitration between Bangladesh and India". He was also the head of the Bangladesh Continental Shelf technical team entrusted with the preparation, documentation and carrying out of seismic survey in the Bay of Bengal and the final submission and presentation to the Commission on the Limits of the Continentals Shelf of the UN. He is currently the President of the Assembly of the International Seabed Authority, Kingston, Jamaica. He was commissioned in the Bangladesh Navy in 1973 and joined the Ministry of Foreign Affairs as Rear Admiral in 2009. On the completion of his graduation and Law of the Sea course from the Royal Naval College, Greenwich, UK, he was awarded with a "Commendation Letter" for his meritorious works. He also received the Indian President's Gold Medal from the Indian Naval Academy in 1974. He has number of publications to his credit and a 2004 book titled, *Bangladesh's Maritime Challenges in the 21st Century.*

Dr Narasimalu Srikanth is with the Energy Research Institute, Nanyang Technological University, Singapore, as a Senior Scientist and Program Director, leading the wind and marine renewable research towards tropical regional needs. He has more than 25 years of industrial experience and earlier he was with the Vestas Windsystems as a senior specialist and Technical Director in the Research and Development department, leading some of the key research activities related to onshore and offshore wind turbines. Prior to these appointments, he also served at ASM Technology Singapore, leading the research and development of semiconductor equipment and with Tata engineering (TELCO), Pune focusing, on heavy machinery design. He received his first masters from IIT Bombay, in machine design and his second masters in materials engineering and his first PhD in the area of mechanical engineering and a second PhD in Technology Management from the National University of Singapore (NUS). Dr Narasimalu has published in more than 100 international journals and has more than 30 patents. His research interests are wind and tidal turbine development, renewable resource mapping and forecasting, structural health monitoring and simulation methods.

Mr Sujay Chohan is the co-founder of Ocean Blue Boating Pvt. Ltd. He is a successful entrepreneur having built and exited two ventures in the advertising/publishing and BPO (IT research/analysis and advisory) businesses. Sujay launched India's first CIO magazine *MIS* and set up a KPO as early as 1997, which was eventually acquired by "Gartner" in 2001. Within a decade of co-founding "Ocean Blue", the company has a number of firsts to its credit and has made some pioneering efforts towards raising the profile of boating, both within India and internationally, commencing with the hugely successful inaugural Mumbai International Boat Show 2007 and the launch of India's first Boating Magazine *India Boating* and building India's first and only marina in Kochi. Immediately prior to founding Ocean Blue, Sujay was the Country Director and Vice President Research at Gartner, responsible for Gartner's Research and Advisory Services and on the Board of Directors of Gartner India. He has travelled extensively globally. He is a frequent speaker at global ICT conferences and is often quoted in both electronic and print media. Sujay holds a Commerce degree from Sydenham College, University of Bombay.

Professor G.S. Roonwal is an honorary visiting professor at the Inter University Accelerator Centre, an autonomous Institute of the UGC/MHRD. He was earlier Professor in Geology and Head of the Department of Geology, University of Delhi. He has participated in the exploration of sea floor manganese nodules and seabed massive sulphides. He has authored 'The Indian Ocean Exploitable Mineral and Petroleum Resources, Springer, Heidelberg', and 'Competition for Seabed Resources in the Indian Ocean (NMF)'. Besides, he is a co-editor of India's Exclusive Economic Zone. Prof Roonwal is a highly decorated geologist, a recipient of the National Mineral Award of the Government of India, and several other awards. He has travelled abroad and worked with distinguished marine geologists overseas. He is also listed in the Marques Who's Who in the World and Who's Who in Asia since several years.

Ms Imali Udeshika Manikarachchi is a lecturer at the Ocean University of Sri Lanka. She received her Bachelor of Science degree in Fisheries and Marine Sciences from the Ruhuna University of Sri Lanka. She worked at the same University as an assistant lecturer for one-and-half years and then joined the Ocean University of Sri Lanka as a lecturer. She received a prestigious Sasakawa Fellowship administrated by the Ocean Policy Research Foundation in Tokyo, to pursue a Master's Degree in Maritime Affairs from the World Maritime University, Sweden. Her area of specialisation extends to Marine Environment Management and her Masters Dissertation was under the title of "Stepping Up from Green Revolution to Blue Economy: A New Paradigm for Sustainable Development in South Asia". Her major academic duties include teaching and conducting research on marine scientific aspects. She has published several articles, papers and abstracts on various topics relevant to Marine Ecology, Marine Management and Blue Economy.

Professor Romeela Mohee is the former Vice-Chancellor, University of Mauritius, Mauritius. She holds an Engineering Degree in Energy and Environment from the Grande Ecoled' Ingénieurs, the Institut National des Sciences Appliquées de Lyon, France. She was awarded PhD from the University of Mauritius in collaboration with the Clemson University, USA, under a Fulbright Scholarship in 1996. She has held the post of National Research Chair in Solid Waste Management at the Tertiary Education Commission and the Mauritius Research Council. She is the first woman Vice-Chancellor of the University of Mauritius.Professor Mohee has received many awards like the Gold Medal for Applied Research, Most Outstanding Young Person Award 2005, Winner of African Union Woman Scientist Regional Award for Science, Technology and Innovation in September 2009, and the Drivers of Change Individual Winner Award from the Southern Africa Trust, the *Mail* and the *Guardian* in October 2015. She has been conferred the award of Commander of the Order of the Star and Key of the Indian Ocean (C.S.K) for her contribution in the field of tertiary education, on the occasion of the National Day 2014. She has recently been conferred the "Officier de l'ordre national du mérite" by the French Republic.

Dr Nirmal Jivan Shah is the Chief Executive of Nature Seychelles, Seychelles. He has a Bachelor's degree in business, an MSc and PhD in biology and ecology. He is a prominent civil society leader in Seychelles and is the Chair of LUNGOS (the umbrella organisation of the NGOs in Seychelles) for six years. He is Chief Executive of Nature Seychelles (Bird Life International Partner in Seychelles). He is also the President's Special Envoy for Environment and Climate Change and Chair of the Seychelles National Environment Advisory Council. He also serves as the Honorary Consul for Mauritius in Seychelles. He was the coordinator of Seychelles' first National Biodiversity Strategy and Action Plan as well as for the Environment Management Plan of Seychelles 2000–2010.Dr Shah is leading the Seychelles' longest running ecotourism programme, based on the Cousin Island, which is considered as a model for small island states. He is a founding member and former President of the Western Indian Ocean Marine Science Association (WIOMSA). He is currently leading the management of a climate change adaptation coral reef restoration project, the world's largest, using the "reef gardening" method.

Inspector General S. Paramesh, TM is the Deputy Director General (Operations and Coastal Security) in the Indian Coast Guard. He joined the Indian Coast Guard on July 17, 1989. The officer, over the last two decades, has served in various capacities in ashore and afloat appointments with distinction. The officer has a professional history, studded with achievements and a proven track record of outstanding and meritorious performance in all the assignments he has held. The Flag officer is specialised in Navigation and Direction and his sea commands include all major ship classes of the ICG, which include the Advanced Offshore Patrol Vessel Samar and the Offshore Patrol Vessel Vishwast. His key staff assignments include Principal Director (Operations) at the Coast Guard Headquarters and Chief Staff Officer (Operations) at the Coast Guard Regional Headquarters (East), Chennai. The Flag Officer is an alumnus of the National Defence College, New Delhi and Defence Services Staff College, Wellington. He is a recipient of the Tatrakshak Medal and was also awarded the Director General Coast Guard Commendation in 2012 and FOCINC (East) commendation in 2009.

Vice Admiral Pradeep Chauhan (Retd), AVSM and Bar, VSM, is an alumnus of Mumbai's Campion School and St Xavier's College, the National Defence Academy, the Defence Services Staff College, the Naval War College, and, the very prestigious National Defence College. He holds a BSc, an MSc and an MPhil degree and has been an Indian Naval officer of great distinction. Admiral Chauhan retired in December 2013, after having done our nation proud for four decades. During this period, he tenanted a number of critical shore-based appointments and, in addition, was entrusted with sea-going command on as many as four occasions, culminating in the command of the aircraft carrier INS *Viraat.* As a Flag Officer, he has enjoyed untrammelled success as the Assistant Chief of the Naval Staff for Foreign Cooperation and Intelligence, the Chief of the Staff of the Western Naval Command. His most satisfying assignment, and perhaps his most meaningful and lasting contribution to the future of our country, has been as the Commandant of the state-of-the-art Indian Naval Academy at Ezhimala, Kerala—a post he held until his retirement. Not one to rest on past laurels, he has remained active even after retirement and is a much sought-after leadership mentor and an outstanding orator. He is also a prolific writer on maritime affairs and leadership issues.

Dr P. Sekhar is the Chairman of Micro Tech Global Foundation. He is a development thinker and author of 60+ books on the national growth of Indian States and UTs, sectors and countries; what they have and what they need for sustainable growth in key sectors. His concept of 'secured governance for techno–economic national growth' offers extraordinary infrastructure development with minimum governmental investment and appreciated by intelligentsia, political heads, bureaucrats and icons in the industry including the late Dr Abdul Kalam, Shri Narendra Modi and Jimmy Delshad (the Smart City man of the USA) and has attracted appreciation in nations like the US, UAE, Tanzania, Kenya, Rwanda, etc. He holds a Doctorate in Solid State Electronics and Nuclear Physics and is associated with governments, industries and universities. He was the first director of the Software Technology Park in India. He is assisting various industries to use innovative technologies with international funding support. More than 100+ technology products have been developed by him, which have been internationally acclaimed.

Dr (Capt) Suresh Bhardwaj is the Resident Director, Marine Training and Research Foundation, at the MASSA Maritime Academy, Chennai. Dr (Capt) Suresh Bhardwaj has 40 years of work experience, which includes five years as Master and in command of various types of ships of the merchant marine and subsequently, 25 years of multidisciplinary shore experience in senior and top management positions spanning the marine industry verticals of commercial operations, consultancy, academia and research. He is on the Board of Advisors of Maritime Academy of Denmark and on the Board of Research of the Indian Maritime University, as well as Adjunct Professor with the Australian Maritime Academy, University of Tasmania, Australia. He is

Co-Chair of Shipping and Logistics Committee of ASSOCHAM, South India and a member of the Local Advisory Board of the NMF, Chennai. He is currently the Resident Director of Maritime Training and Research Foundation that is promoted by MASSA (Maritime Association of Ship Managers, Ship owners and Agents) and heads its Maritime Academy. He has pioneered the concept and implementation of 'Maritime Cluster' and 'Knowledge Partner' in 2014 with initiatives at the Gujarat Maritime Board. In this presentation, he reinforces the 'Maritime Cluster' concept and outlines how the 'Cluster' environment will boost the entrepreneur and start-up enterprises across the various verticals and segments of the shipping industry and in various product forms of hardware, software or service.

Ms Vasudha Chawla is working as a Research Associate at the National Maritime Foundation, New Delhi. She has completed her Masters in Defence and Strategic Studies from Panjab University, Chandigarh and a Diploma for Graduates in Management from University of London International Programmes, based in New Delhi. Her research interests include India's developments with respect to the Blue Economy, inland waterways and connectivity infrastructure.

Index